I0798841

IN GOOD FORM

ARGUING FOR EPISTEMIC
NORMS OF CREDENCE

Leszek Wroński

IN GOOD FORM

ARGUING FOR EPISTEMIC NORMS OF CREDENCE

Jagiellonian University Press

REVIEWER
dr hab. Rafał Urbaniak, prof. UG

COVER DESIGN
Agnieszka Winciorek

This book has been published with the financial support of the Institute of Philosophy at the Jagiellonian University in Kraków.

© Copyright by Leszek Wroński & Jagiellonian University Press
First edition, Kraków 2018
All rights reserved

No part of this book may be reprinted or reproduced or utilized in any form or by any eletronic, mechanical, or other means, now known or hereafter invented, including photocopying and recording, or in any information storage or retrieval system, without permission in writing from the publishers.

ISBN 978-83-233-4457-5
ISBN 978-83-233-9822-6 (e-book)

JAGIELLONIAN UNIVERSITY PRESS
www.wuj.pl

Jagiellonian University Press
Editorial Offices: Michałowskiego 9/2, 31-126 Krakow
Phone: +48 12 663 23 80, +48 12 663 23 82, Fax: +48 12 663 23 83
Distribution: Phone: +48 12 631 01 97, Fax: +48 12 631 01 98
Cell Phone: + 48 506 006 674, e-mail: sprzedaz@wuj.pl
Bank: PEKAO SA, IBAN PL 80 1240 4722 1111 0000 4856 3325

To my dearest wife Marta

How could we express our ideas without numbers?
– C.P.E. Bach, *Versuch über die wahre Art das Clavier zu spielen*, part II, chapter 1, pt. 4[1]

[1] Translation mine. Yes, he was talking about figured bass; see the introduction to this volume.

Contents

Introduction: formal philosophy as figured bass

In this volume I attempt to contribute to a few philosophical discussions. I try to clarify a few details relating to so-called "Dutch Book arguments"; I present fully written out proofs of a few seemingly neglected *theorems* of that kind and argue that the most popular *arguments* based on them do not hold water. My take on the "undermining" problem that supposedly plagues Humean theories of chance and what the "Best System" can be expected to offer a Humean is presented next. I then discuss some issues regarding higher-order probabilities, for example adding propositions specifying the probability of some events to a probability space containing those events, and I present a related construction and prove that it works.

In the second part of the book I discuss some ways of measuring the value of one's credal state and assessing the relationship between two credal states which could give rise to epistemic norms like Conditionalization and its variants. As for the second issue, I will argue that Inverse Relative Entropy (IRE) is the way to go for a variety of reasons. I prove a few theorems concerning the behavior of IRE and a few competing methods across different belief update problems. Then I discuss the so-called "elimination counterexamples" which form the basis of arguments against using the Brier Score as a measure of epistemic inaccuracy of an agent's credal state, arguing, by way of examples and theorems, that we need not be worried by them too much. I also strengthen some arguments *for* using the Brier Score found in the literature, as well as present some related open problems.

The general topic to which the book tries to contribute is that of epistemic norms for degrees of belief and how to argue for them. One of the foremost norms of such a kind, Probabilism, and probably the best-known argument for it, the Dutch Book Argument, are discussed in Chapter 2, which is completely self-contained. Another famous norm, the Principal Principle, is the topic of Chapter 3. Chapters 1 and 4 concern the ways in which arguments are conducted in formal epistemology. Chapters 5 and 6 aim to present two faces of a position I would like to propose, that of *formal justificational pluralism*: it is possible that methods which may be used when arguing for synchronic norms are different from those which may be used when arguing for diachronic norms.

*

But, apart from a genuine drive to offer a positive contribution to the philosophical community, this book partially arises out of frustration.

Back in 2008 I was writing my PhD thesis on new (then) and—I thought—exciting developments around some formal renderings of Reichenbach's Common Cause Principle. As you can expect from the connection to Reichenbach, the topic sometimes touches the realm of (philosophy of) physics, and so papers in that field usually contain at least one theorem, the custom being that the proof is relegated to some appendix. So, I started reading papers published in *Philosophy of Physics*, *Philosophy of Science*, *The British Journal for the Philosophy of Science* and so on, meticulously going through all the proofs. I am unable to read papers in formal philosophy differently—if I skip a proof, I feel that I have not understood the theorem. This of course makes for slow reading, but I am all for some additional time investment if it brings more clarity.

At a certain point I started to have some trouble with a proof from one of the papers published in a respectable journal whose main topic was philosophy of science. There was a proof by cases and it seemed to me that one case was missing. I set out on the task of filling in the blanks; to my surprise I discovered a counterexample to the theorem, the proof of which I was trying to complete. That is, the theorem—one of the main points, if not *the* main point made in the paper—turned out to be false. My first thought was, of course, "the world needs to hear about this!". (I got a paper out of the whole thing.) However, I then asked myself: has no one but me read that proof? We all make mathematical mistakes, of course, but catching them is a job for the reviewers. Moreover, the paper had been in circulation for a few years, and not a single author had commented on the incorrect argument. I asked around a summer school I was attending at that time and made some inquiries during a conference or two among fellow PhD students: does anyone read proofs? My interlocutors almost invariably responded: "It's just you, Leszek. Who has time for that?". Well, I already then feared I could not shake that habit. After defending the PhD and publishing a few papers and a book to the calm indifference of the public I decided that I should change fields: my future research would belong to the wildly flourishing field of formal epistemology. The fact that the circle of researchers involved would be much wider *surely* meant better peer control, and theorems a reader could trust to be true!

Boy, was I wrong. At the moment of writing these words it is clear to me that respectable journals like *Noûs* or *Philosophical Studies* have published papers whose formal parts had not been seriously read by anyone. I do not wish to criticize the authors—it is evident that each paper involved contains an interesting idea (in some cases I will elaborate on this in subsequent chapters). However, frequently the "theorems", when seriously considered, do not make sense *at all*, or, if supposed to be true, have absurd or outlandish consequences. What is more, it seems that formal philosophy is plagued with incorrect citations: sometimes authors quote other authors incorrectly, and then base their arguments at least partially on those quotes. (I am not accusing anyone of ill will; these are surely all honest mistakes.) This happens, as I will substantiate later on, not just to lowly Eastern European authors like myself who are usually not cited at all, but to true stars like David Lewis and well-known living philosophers. I try to keep up with the literature on a few strands of philosophy and musicology; in my experience the situation with citations and, more important, incorrect usage of formal methods is nowhere as bad as in formal epistemology.

I propose we do our best to change this situation.

If formal epistemology, and formal philosophy in general, is to be taken seriously, that is, if we who dabble in it want to take ourselves seriously, if we want to be able to honestly think we are doing a good job, we need to get our affairs in order. Most importantly, let us not pretend we are proving stuff when in fact we are not. Let us not call a "theorem" something which, if we look into our hearts, is really a poetic sketch of a somewhat vague, philosophically interesting idea. If the idea is good, "formalizing" it will not improve the situation: in fact, it may hamper understanding. If the idea is not good, instead of hiding it behind layers of fake formalism, we should not write about it. Pretending to be formal about stuff which is best talked about in an informal way is tantamount to Sci-Φ, that is, a combination of philosophy, science, and fiction—philosophy pretending to be science. I propose, then, that we adhere to the following:

The Prime Directive: Do not overformalize. But if you formalize, do it properly.

If the Prime Directive seems to be a pale variant of the Aristotelian thesis that "Our discussion will be adequate if it has as much clearness as the subject-matter admits of, for precision is not to be sought for alike in all discussions",[2] then so be it. A greater part of the book can be seen to be an elaboration of this thesis. I believe in philosophy there are less fruitful endeavors. The current one offers the reader a good deal of pain and exhaustion, with a vague promise of leaving the field a somewhat hardened mathematical philosophy combatant.

*

It is not a coincidence that the motto of this book comes from a famous volume on the art of playing keyboard instruments, and in particular on the writing and performance of figured bass.

---

[2] *Nicomachean Ethics*, Book I, translated by W.D. Ross.

In my opinion, in formal philosophy—that is, philosophy which does not shun mathematics, logic and the sciences, and which attempts to use formal methods itself—the formal aspects should play the same role figured bass plays in XVIIIth-century music. Figured bass is a bass melody with a sequence of (tuples of) natural numbers and other symbols above it, like this[3]:

The numbers code the harmony which is to be heard at a given moment; various optimizational conventions were adopted around Europe so that the lack of certain numbers at certain places is also meaningful. Figured bass conveys information to players of instruments like the lute, harpsichord, or the organ—ones that are able to emit the sound of many notes at the same time. The coded information is that of harmony, but a skilled player will typically realize it so that a melody becomes audible which bears a resemblance to what's going on elsewhere in the music. (The information may also be invaluable to a musicologist trying to reconstruct a partially destroyed work.) There are many ways a figured bass may be realized, but two points are essential: that there is a clear concept of a "mistake", and that the figured bass itself does not make or break the work. For example, a piece with a lousy figured bass can be saved by virtuoso writing for solo violin.[4]

Consider two of the infinitely many ways of realizing the figured bass presented above:

---

[3] Here and elsewhere, music is engraved using Lilypond 2.19.48, www.lilypond.org.

[4] See, e.g., the violin sonatas by Schreyvogel from the DUX0968 recording "Music in Dresden in the times of Augustus II The Strong".

This is a mundane, chordal realization of a figured bass. It fills in the whole harmony as required and will not divert the listener's attention away from other elements of the music. However, on its own it is rather dull.

This "melodic" approach does not offer the whole harmony but presents a musical idea which might fit with other motives in the work, adding to the listener's sense of cohesion.

What the two examples have in common is that they indeed adhere to the figured bass: on each beat there is no sound which does not follow from the "code". Both of them are "correct". It is my opinion that just like how a figured bass does not dictate a unique way of playing a given piece, the entirety of a paper in formal philosophy should not be governed by the formal results therein. However, just as any "correct" realization of a figured bass is careful not to break any of the rules, we should make sure that the more philosophical parts of our work really fit with the formal ones. Fundamentally, just as anyone preparing a figured bass line (not the performer realizing it, but the composer *writing*, i.e., coding, it) needs to adhere to a set of rules (see C.P.E. Bach's *Versuch* for an extended discussion), we who attempt to use formal methods in philosophy should make sure we are actually using them; a composer writing random numbers around his bass lines would be ridiculed (or at least misunderstood) by performers, and we should be humble enough to expect nothing else if we, say, do not define our structures properly (see Chapter 1 below for examples of this and other related phenomena).

(Of course, in music one needn't use figured bass at all, just as in philosophy one can do much without any formalism. But this is not a book for people who are interested only in *that* sort of philosophy.)

*

Acknowledgments

I would like to thank Michał Tomasz Godziszewski, Richard Pettigrew, Tomasz Placek and Andrzej Wroński for commenting on chapters of this book.

I developed much of the material as a member of the Budapest–Kraków Research Group on Probability, Causality and Determinism. I thank all the members of the group and speakers as well as guests of the group's workshops for interesting discussions. I have benefitted greatly from co-authoring papers with Balázs Gyenis and Michał Tomasz Godziszewski.

I would like to thank the following people for various bits of help they have graciously given me when I was writing the book, in alphabetical order: Gergei Bana, Branden Fitelson, Michał Tomasz Godziszewski, Balázs Gyenis, Zalán Gyenis, Remco Heesen, Ronnie Hermens, Richard Pettigrew, Tomasz Placek, Jan-Willem Romeijn, Jan Szwagierczak, Rafał Urbaniak, Jan Woleński.

*

Funding

The research was partially financed by my Narodowe Centrum Nauki SONATA grant "Epistemic inaccuracy and degrees of belief" ("Nietrafność epistemiczna a stopnie przekonań", UMO-2015/17/D/HS1/01912).

*

Relation to other publications

Sections 5.1–5.4 are based on a paper I published in *Ergo* in 2016 to which I—due to the beautiful nature of the Author Agreement of that fabulous Open Access journal—retain the copyright.

Otherwise, if a fact or theorem is lifted from material I have published elsewhere, the source is always given.

Part I

Some basic issues in formal epistemology

Chapter 1

A few motivational examples

In this chapter I will describe a few examples of the problems which, I think, have troubled the beautiful field of formal epistemology throughout recent decades. I feel a bit uneasy about this—let it be said, then, that I firmly believe that all the authors mentioned later in this chapter are way smarter and more philosophically competent than me, while also quite possibly possessing a better command of mathematics. It is just all the more unfortunate that their great work is blemished by the imperfections I wish to point out, hopefully to the benefit to those of us who wish to contribute to the field. Since I do not want to come out as needlessly negative, I will restrict myself to a single example of each type of the problems.

1.1. "Proofs" which are not really proofs

The classical Dutch Book Argument (DBA)—which we will discuss in detail in the next chapter—aims to establish some constraints on rational degrees of belief, or credences, based on a formal result describing a class of agents vulnerable to a "sure-loss" betting setup, a "Dutch Book" (see, e.g., Vineberg (2016); the details of this are philosophically controversial and thus fleshed out in innumerable ways, including some which seem to be decisively less pragmatic). The result at the heart of the argument is the so-called Dutch Book Theorem (DBT), which, strictly speaking, concerns the agent's betting quotients (or odds, etc.), the relationship between which and the agent's degrees of belief may not be trivial if you are not a strict operationalist like de Finetti (we will also talk about this below).

The Theorem is a famous foundational result of formal epistemology, usually attributed e.g. to Ramsey (see, e.g., the 1931 paper, which combines a few of his earlier writings). It describes the necessary and sufficient conditions for the betting quotients being coherent, that is, such that they exclude the possibility of the aforementioned Dutch Book. The reason for its popularity was probably the remarkable simplicity of the conditions: it turns out that they correspond to the (now) classical axioms of probability. That is, as Gillies puts it in his 2000 monograph *Philosophical theories of probability*, a fantastic and widely cited introduction to the field (especially the propensity theories):

> A set of betting quotients is coherent if and only if they satisfy the axioms of probability. (Gillies (2000), Chapter 4)

To prove the theorem, one proceeds in two directions. First, that for each probability axiom, its violation by the betting quotients implies incoherence. Recipes for this can be concisely stated and are easy to illustrate, which is why most authors writing on the topic only give the proof of this direction of the DBT. The second one, called also the "Converse Dutch Book Theorem", is—despite what you might have heard (I certainly did during conferences, and I'm not particularly well-travelled)—much trickier and it seems to me that it is no surprise that its proof has been "surprisingly neglected" (Hájek (2008)); we will devote a serious chunk of the next Chapter to its careful presentation and analysis. What one needs to show is that the fact that betting quotients satisfy *all* the probability axioms guarantees coherence; that is, there is no "sure-loss" scenario for the agent. Let me stress here that all the axioms are needed for this: even without knowing any details of the arguments, the reader has already received the information from the "first" direction of the Theorem that the violation of even a single axiom leads to incoherence.

The way Gillies (2000) presents the proof of the Converse DBT (p. 61) has to then come as a surprise. The author claims he shows, for each axiom, that it by itself guarantees coherence. The reader should at this point immediately see that this cannot work: again, the fact that one of the probability axioms holds does not guarantee that some other does not, which leads to incoherence! (Apologies for reiterating the point, but my excuse is that since this error has been present, apparently unnoticed, in so many reprints of Gillies' book, it is at least not *evidently* evident.) An examination of the argument, I submit, has to leave one perplexed: the "proof" given by Gillies does not establish the Converse DBT, and the mistake is not hidden deep in the argument as if it were a reward for an enterprising reader to find, but is instead an immediate structural failure.

What is more, the proof is supposed to be an expanded version of the argument from de Finetti (1937/1964). (Gillies is an expert on de Finetti, see, e.g., Gillies (1972).) In the next chapter we will closely study three arguments for the Converse DBT, including de Finetti's, to see what this is all about.

1.2. Arguing using undefined structures

1.2.1. Undefined structures, implicit quantification: the Qualified Reflection Principle

Continuing the discussion of alleged proofs, but attempting to showcase a different argumentational flaw, I will now present a case—published in a top philosophical journal (Briggs (2009))—in which lack of definitions of the structures involved combined with the omission of explicit quantification leads to a "Theorem" which not only has arguably inconsistent assumptions, but also possesses an incorrect "proof". As was the case with Gillies's book, I do not want the reader to think I have a negative opinion of Briggs' paper—to the contrary, I think it is of high philosophical merit and the idea of Distorted Reflection that is introduced there should be closely studied. The message should rather be that if the field of mathematical philosophy is to be taken seriously, we need to pay more attention to the math actually used in the papers.

The usual interpretations of the various variants of the norm of Reflection (introduced in van Fraassen (1984)) deal with current credences of a rational agent conditional on her future credences. The argument I want to analyze is meant to establish the so-called "Qualified Reflection" principle (Briggs (2009), p. 69), the formalism of which uses two symbols for credence functions: Cr_0 for the credence of some agent at time t_0, and Cr_1 for the credence of that agent at a later time t_1. (No structure is defined as the domain of either Cr_0 or Cr_1; I proceed with the typical assumption that in both cases it is the same algebra of subsets of some base set.) First, we need three "idealizing assumptions" (all quotes are from p. 69 in Briggs (2009)):

1. "I will assume the agent is a perfect introspecter—in other words, that $Cr_0(Cr_0(A|B) = r) = 1$ if and only if $Cr_0(A|B) = r$". I gather that universal quantification over A, B and r is meant, together with a proviso to the effect that $Cr_0(B) > 0$—otherwise the conditional credence will not be defined. (We should assume that both Cr_0 and Cr_1 are probability functions.)

2. "I will assume that the agent's possible evidence propositions—that is, the propositions that might represent the totality of what the agent learns between t_0 and t_1—form a partition $\{B_1, B_2, \ldots B_n\}$". Since there is no further comment on this, I gather that it is a partition of the base set on an algebra of subsets of which the functions Cr_0 and Cr_1 are defined.

3. "I will assume that all agents can reasonably be certain that conditionalization is the right updating procedure".

Then, the norm of Qualified Reflection is given as the following:

Qualified Reflection: $Cr_0(A|Cr_1(A) = r) = r$, provided that for all $B \in \{B_1, B_2, \ldots B_n\}$,

i. $Cr_0(Cr_0(A|B) = Cr_1(A|B)) = 1$ and

ii. $Cr_0(B|Cr_1(B) = 1) = 1$. (Briggs (2009), p. 69)

In the statement of the norm, B is bound, and I gather that universal quantification is meant over A and r.

According to Briggs, the norm of Qualified Reflection, given the three idealizing assumptions above, follows "from the Kolmogorov axioms". Assume, however, that some proposition A is given. It follows from condition i. that for all $B \in \{B_1, B_2, \dots B_n\}$ the conditional credence $Cr_1(A|B)$ is defined, that is, that $Cr_1(B) > 0$, and so, if $n \geqslant 2$, that for all $B \in \{B_1, B_2, \dots B_n\}$ $Cr_1(B) < 1$. We have thus, thanks to assumption 2, established that the agent will not, between times t_0 and t_1, update her credence function using conditionalization, since that would require that for one of the B_i's, $Cr_1(B_i)$ should be equal to 1. This is not an outright logical contradiction with assumption 3 (such a contradiction will be hard to find anyway, since that assumption is not given a formal statement), but clearly something is wrong here: we require (in 3) that the agent be certain that conditionalization is the right updating procedure, and yet the norm we wish to impose makes it impossible for that agent to conditionalize. Isn't this unfair towards the agent?

What is more, the proof given by Briggs of the fact that Qualified Reflection "follows from the Kolmogorov axioms" on p. 69 of the paper is incorrect, and for a reason which it may be instructive to inspect. Namely, the argument requires divisions by results of various sums taken over the sets $\{B \in \{B_1, B_2, \dots B_n\} \mid Cr_1(A|B) = r\}$ and $\{B \in \{B_1, B_2, \dots B_n\} \mid Cr_0(A|B) = r\}$ (these sets are defined by the choice of r). But there is no guarantee in general that these sets are nonempty, and so no guarantee that the sums are not equal to zero! In fact, it is immediate that given a proposition A, each of these sets will be empty for all but finitely many choices of r. So, if the proof is supposed to go through, the norm of Qualified Reflection needs to be additionally constrained; I do not see a convenient and nontrivial way of doing this.

These issues could have been avoided if the structures employed had actually been defined and the quantification had been made explicit. However, defining structures appropriate for rigorous argumentation regarding credences about credences or credences about chances is no mean feat and is not usually attempted. I will now point out an example of a different problem to which this situation leads.

1.2.2. Conclusions about structures which are undefined: arguing about chance propositions

Another famous norm which, taken seriously, requires one to consider probabilities of probabilities, that is, higher order probabilities, is D. Lewis's "Principal Principle", some issues regarding which we will discuss in Chapter 3. It refers to rational credences about chances, so if rational credences are probabilities (as argued for example in the DBA) and chances are probabilities also, higher order probabilities are certainly involved. Done properly, these probabilities would be defined on an algebra of propositions, some of which would be "chance propositions", that is, they would be propositions that the probability of some proposition equals some value. Such decidedly nontrivial constructions were the topic of for example the seminal Gaifman (1988) (see Chapter 4 below); however, this is not usually done. Typically, authors are just content with slapping a "chance proposition" label on some seemingly arbitrarily chosen propositions, proving some mathematical statement, and inferring some philosophical conclusions about chance propositions.

An interesting example of this can be found in Hawthorne et al. (2017), a paper which intends to deliver a potent philosophical message connecting two famous principles: the Principal Principle and the Principle of Indifference. For the time being, let me notice that (as discussed in Gyenis and Wroński (2017) the main result of the paper is equivalent to the following:

> **Proposition 2 from Hawthorne et al. (2017)**: Let $P(A|X) = P(A|XE) = P(A|FXE) = P(A|(A \leftrightarrow F)XE)$. Then from $x \neq 0$, $x \neq 1$ and from the Principal Principle $P(A|X) = x$ it follows that $P(F|XE) = 1/2$.

on the assumption that, quote, X "says that the chance at time t of proposition A is x and E is any proposition that is compatible with X and admissible at time t". Forget about "compatibility" and "admissibility" for now. But do notice that the status of X as a chance proposition is set only by the vague "says that (...)" in the metalanguage, with no bearing on the actual mathematics, since the result (which in my opinion is correct) follows only from the specified independence assumptions.

However, the authors take it to be a discovery that the "line of argument does not depend on the structure of the proposition X". I cannot help but be quite puzzled by this. It *cannot* depend, because the structure *is not there*—it was not defined anywhere! The rigorous mathematical argument is fully oblivious to what labels we put on the propositions using the natural metalanguage; if we wanted, we could call them "ethical propositions", "colour propositions" or what have you. For the argument to uncover something other than a relationship between a specific type of statement regarding the conditional probability of 1/2 and a series of independence assumptions, and rather to involve propositions about probabilities of other propositions, the relationship between a proposition about the chance of a proposition A and the proposition A itself needs to be a feature of the structure on which the probability functions are defined. (One way of looking at what Gaifman (1988) does is to see it as formalizing this notion of "aboutness".)

1.3. Not *really* reading what one cites

When one *does* want to try to get more serious about higher order probabilities, it is customary to cite Gaifman's seminal paper "A Theory of Higher Order Probabilities" and offer a few words of commentary. For example, Dziurosz-Serafinowicz (2016), in Section 1.4.1, writes that according to Gaifman "we can enlarge the original set of propositions over which a probability function is defined by adding propositions about probabilities that this probability function assigns over propositions in the original set", which is a concise way of describing exactly what the paper is about. However, most authors, like Dziurosz-Serafinowicz himself later in the cited work, are mostly interested in probabilities of "chance propositions", that is, propositions that the *whole* chance function is this or that, and not just propositions that the chance of some event equals some value. That chance propositions do belong to the structures defined by Gaifman is trivial in countable cases, but requires an argument in others, an argument I have never seen anyone make.

What is more, authors who aim to produce the Gaifman-inspired structures typically do not check whether the objects they construct indeed satisfy Gaifman's requirements. After reading many papers allegedly employing Gaifman's techniques and talking to a few of the authors at some conferences, I began to suspect that actual knowledge of what Gaifman wrote is not that common.

These suspicions seem to have been confirmed by the appearance in 2016 of a "Sourcebook" entitled "Readings in Formal Epistemology" (Arló-Costa et al. (2016)), with the honorable aim of presenting "38 classic texts in formal epistemology", including the one by Gaifman. Note that three of the editors are towering figures in epistemology. Now, Gaifman actually published *two* papers with the same title, in 1986 and 1988. The later version is in some respects superior to the earlier one. Not only does it better connect the issue to existing work, offering about twice as many references and, for example, slightly extending the connection to modal logic at the very end, but also it contains a worked-out Dutch-Book-style argument which is only hinted at in the 1986 paper, making it significantly more interesting from the philosophical point of view. It is also more clearly typeset. It seems to me, then, that currently the only reasonable motivation for consulting the 1986 paper would be an interest in the history of the evolution of Gaifman's thought.

On the assumption that one is interested in the actual content of the Paper, it is therefore quite surprising to notice that the 2016 "Sourcebook" contains the *older* version of Gaifman's work; that is, none of the five (!) editors spotted that they were offering newcomers to the field a text which had been improved by the author himself in a publication almost three decades before.[1] On the other hand, on the assumption that the paper is just something one should cite to show erudition when writing about higher order probability spaces, the situation is perfectly understandable.

[1] Somewhat grotesquely, it seems that someone was actually hired to LaTeX an old version of the Gaifman paper even though the later one is properly typeset.

1.4. Misquote a classic, no one notices

In modern formal epistemology it might also happen that quotations are not accurate. It is of course normal to change the notation when quoting a formula so that it fits one's chosen formalism. It is another thing, though, when similar changes inadvertently modify the philosophical argument itself. This is doubly unfortunate if the issues under discussion are so notoriously muddled as those of "admissibility" and the "Principal Principle"—to which we will return in Chapter 3—which seems to be what happened in the otherwise magnificent paper by Ismael (2008). Consider this quote, allegedly from (Lewis (1994), p. 485), which I'm reproducing here with Ismael's notation, with the symbol "Ch" denoting "chance", and "$\mathbf{PP}_{\mathbf{orig}}$" denoting the "Principal Principle", which we will not talk about until Chapter 3:

> [O]ur problem, where F is an unactualized future that would undermine the actual chances given by E is that Ch(F/E) = 0, because F and E are inconsistent, but Ch(F/E)≠0 by $\mathbf{PP}_{\mathbf{orig}}$ because E specifies that F has some present non-zero chance of coming about. (Ismael (2008), p. 295)

And now compare it with the Lewis original:

> Our problem, where F is an unactualized future that would undermine the actual present chances given by E, is that C(F/E) = 0 because F and E are inconsistent, but C(F/E)≠0 by the Principal Principle because E specifies that F has non-zero chance of coming about. (Lewis (1994), p. 485)

Notice the differences. One puzzling thing is Ismael's interesting relocation of the word "present", which in itself could spark, I am sure, a few pages worth of discussion, a temptation I shall resist. More important, though, is the seemingly innocuous transformation of Lewis's "C" into Ismael's "Ch"; the change is not innocent at all, since the symbol signifies the *credence* function for Lewis, and the *chance* function for Ismael, so if chance–credence principles are the topic, this can only introduce confusion.

Ismael by mistake attributes to Lewis a passage which is not trivially wrong[2]; in fact, it suggests a certain direction of reasoning using the Principal Principle which at first glance might be novel (though not Lewisian, and ultimately mistaken). Even though, as it seems to me, the portion of Ismael's paper discussing the quote suffers from the quote itself being inaccurate, the author proceeds to introduce a way of connecting credence (and thus ignorance) with chance, which has many things speaking for it (The "General Recipe", p. 298). It is quite interesting to see that in his critique of it, Pettigrew (2015) gives it a form (labelled as 'IP', p. 179) in which I at least cannot recognize the original (for starters, conditional chances are introduced which are missing in Ismael's formulation). This, however, has led to some fruitful philosophical discussion (see the response in Ismael (2015)).

I will try to contribute to the topic in Chapter 3, where I will propose a way of understanding what the so called Best System analysis can offer so that the problem alluded to in the two quotes disappears.

---

[2] That said, there will surely be people who will claim the mistake was trivial. Well, none of the editors and reviewers for *Noûs* found it, apparently.

Chapter 2

The Dutch Book Theorem and Argument

In this chapter I will present the Dutch Book Argument (DBA) for Probabilism, dealing with rational degrees of belief, and the Dutch Book Theorem (regarding Probabilism) (DBT), which deals with numbers intimately related to betting, that is, betting odds or betting quotients. In the first section I will attempt to elucidate some technicalities regarding the DBT, in particular its converse direction. The relation between the DBA and DBT will be the topic of the second section.

Note that we will only be concerned with finite structures and so the only version of the additivity axiom of interest to us is the finite one. I encourage any reader interested in the issues regarding infinite sets of bets and so on to consult Arntzenius et al. (2004).

Throughout this chapter assume (even though we will repeat this for clarity from time to time) that for some nonempty set W, the Boolean algebra of propositions $\mathcal{F} = \mathcal{P}(W)$ is given (with the usual set-theoretic operations), with $\top$ being the top element (the tautological proposition). For future reference, let $\mathrm{At}_{\mathcal{F}}$ be the set of atoms of $\mathcal{F}$, that is, the singletons of elements of W; these are the "atomic propositions", corresponding to the most fine-grained possibilities.

2.1. The Dutch Book Theorem and why it holds

Ideally, to completely demarcate one from another, it would be possible to go "mathematics first, philosophy second". However, the mathematics will only make real sense, that is, will offer a clue as to why we are proving the things we are, if an interpretation is included. Therefore, some discussion of betting, a description of the framework chosen to display the mathematics behind the DBT, belongs in this first section too.

Definition 1 (Finitely additive probability function (measure), space): Assume that $W \neq \emptyset$ and $\mathcal{F} = \mathcal{P}(W)$. A function $p : \mathcal{F} \to \mathbb{R}$ is called a *finitely additive probability function* (or *measure*) iff the following three axioms are satisfied:

(A1) for any $A \in \mathcal{F}$, $p(A) \geqslant 0$ (nonnegativity);

(A2) $p(\top) = 1$ (normalization);

(A3) for any $A, B \in \mathcal{F}$ such that $A \cap B = \emptyset$, $p(A \cup B) = p(A) + p(B)$ (finite additivity).

If p is a finitely additive probability function, then the triple $\langle W, \mathcal{F}, p \rangle$ is a (finite) *probability space*, with W frequently called a *sample space*.

(This is a simplification of the usual mathematical definition of a probability space; there, the $\mathcal{F}$ need not be the full power set of W, but may be any σ-algebra of its subsets, and countable instead of finite additivity is called for in the axioms. This will have no bearing on the arguments below.)

This section is devoted to the mathematics of the DBT. To study it we need to introduce a certain class of numbers related to betting: the given agent's *betting quotients*. The crucial thing is the relationship between a betting quotient and the payoff matrix of a bet. Many authors, including de Finetti and Kemeny, wrote about them as if they resulted from actual betting situations in which an agent really "offers to pay (...)" (Kemeny (1955), p. 263). This gives the analysis more rhetorical bite and offers a pleasing illusion of empiricism, but introduces unnecessary problems; in Kemeny's case, for example, there is implicit universal quantification over the stakes, while in reality it would matter whether hundreds or millions of dollars potentially changed hands. It seems to me that the author was simply not concerned with this, but with the beautiful mathematical result whose goal was to furnish a justification of the above three probability axioms. I will, then, continue to write about the bets an agent "buys" or "sells", etc., but advise the reader to take any mention of "gain" or "loss" with a grain of salt (for example since—obviously—to avoid loss any agent could abstain from betting); we will analyze this issue more closely in Section 2.2. For now, feel free to imagine real people exchanging real money of arbitrary value; also, I will write of "q's profit" when q is a betting quotient function, and the intended interpretation is that the function codes some agent's attitude towards bets.

That said, you will probably need to throw the knowledge you have about actual books, bets and stakes out of the window. The intended framework, that is, the framework which in my opinion makes the mathematics simplest, has the following view in mind. Assume S is some sum of money in agent B's pocket. An agent A *buys the bet for* some proposition E by giving the person B some other sum of money s, on the condition that:

- if E turns out to be false, that money is lost to him, so s stays in B's pocket and A ends up with a "profit" of $-s$;
- if E turns out to be true, B pays S to A but does not return the s, so A ends up with a profit of $S - s$.

In this picture the stake, therefore, is not lumped together from the contributions of two agents; it sits wholly in the pocket of one of them. An agent's betting quotient function determines the s: given the S and a proposition E, it returns a fraction of S for which the agent is willing to buy the bet for E.

In the mathematics below the stakes can be negative; note that nothing of essence changes in the picture just sketched, apart from the two agents exchanging places.

If you take a look at a few randomly chosen texts which try to introduce some version of the DBT, typically both "buying" and "selling" bets both "for" or "against" events is mentioned. At the risk of belaboring the point, I will now attempt to clarify these issues since in my experience they are frequently quite convoluted. Fortunately, it will turn out that only one of the four logical combinations, for example "buying" bets "for" events, will suffice for us to say everything that we want.

A *betting quotient* function $q : \mathcal{F} \to \mathbb{R}$ attaches real numbers to propositions; abstractly, it is indistinguishable from a credence (a degree of belief function), which we will start discussing in Section 2.2. For an $E \in \mathcal{F}$, we will frequently write q_E instead of $q(E)$. A bet *for* E is one that pays some stake S if E is true and 0 otherwise. The quotient q_E is the portion of the stake the agent is willing to pay (or considers to be a fair price, etc.) for the bet for E (and is to be independent of the stake). With such a betting quotient for E, the agent will also accept the payment $q_E S$ for a bet for E which will have her pay out S to the buyer if E is true; in other words, the agent *sells* the bet for E for $q_E S$.

A bet *against* E is one that pays S if $\neg$E is true and 0 otherwise. If the agent's betting quotient for E is q_E, then she is willing to pay $(1 - q_E)S$ for the bet against E, and to sell the bet for the same.[1]

In other words, the quotient q_E determines the following payoff table for bets bought by the agent (columns correspond to states of affairs, rows to types of bet, and each entry is the profit of the agent buying a bet of a given type if the given state of affairs obtains):

BOUGHT	E	$\neg E$
bet for E	$(1 - q_E)S$	$-q_E S$
bet against E	$-(1 - q_E)S$	$q_E S$

As for the bet sold by the agent, the situation is as follows (each entry is now the profit of the agent *selling* the bet of the given type if the given state of affairs obtains):

SOLD	E	$\neg E$
bet for E	$-(1 - q_E)S$	$q_E S$
bet against E	$(1 - q_E)S$	$-q_E S$

[1] If you would like to raise some concerns that we are presuming too much about betting quotients at this point, please wait for Section 2.2.

Observe that a sold bet against E is identical to a bought bet for E (with the same stake). (Assuming that the criterion of identity for bets is the identity of all the entries in the payoff tables; the actual words possibly used during the betting and so on are irrelevant.) A sold bet for E is identical to a bought bet against E (with the same stake). Therefore, we do not need to talk about bets which are sold at all: we can restrict our attention only to bets which are bought by the agent.

What's more, notice that a bet bought against E with stake S is identical to a sold bet for E with the stake $-S$. Therefore, if we allow negative stakes (and we do), we can talk exclusively about bets bought for propositions. This means we will only need to use the following profit table:

BOUGHT	E	$\neg E$
bet for E	$(1 - q_E)S$	$-q_E S$

If instead of talking about betting quotients, one prefers to speak about "odds", keeping with the common usage of the word we can define "odds on E" as the fraction $q_E/1-q_E$. (That is, the betting quotient $q_E = 3/4$ translates into 3:1 odds on E. One can see then why the word "quotient" is in the name "betting quotient".) This, however, forbids us from talking about betting quotient functions which obtain value 1, at least before we make some decision about such cases; *prima facie*, our arguments will be less general. We will therefore not employ odds in this section and stick with quotients otherwise.

Definition 2 (Bet): A *bet* is a pair $\langle E, S\rangle$, where $E \in \mathcal{F}$ and $S \in \mathbb{R}$.

In the bet $\langle E, S\rangle$, E is the event for which the bet is bought, and S is the stake. If $\mathcal{B}$ is a set of bets, we will write "$E \in \mathcal{B}$" with the intended meaning "E is one of the events bet upon", that is, as a shortcut for $\exists\langle B_i, S_i\rangle \in \mathcal{B}\ B_i = E$.

Note that the set of bets $\{\langle E, S_1\rangle, \langle E, S_2\rangle\}$ is equivalent payoff-wise to the singleton $\{\langle E, S_1 + S_2\rangle\}$. We will then assume that for any set of bets $\mathcal{B}$ and for any proposition E, there is at most one pair in $\mathcal{B}$ such that E is its first element.

Definition 3 (Full set of bets): A set of bets $\mathcal{B}$ is called a *full set of bets* if for every $E \in \mathcal{F}$ it is the case that $E \in \mathcal{B}$.

That is, a full set of bets contains bets on all propositions in $\mathcal{F}$.

As already mentioned, the singletons of elements of W, that is, the elements of $At_{\mathcal{F}}$, correspond to the most fine-grained possibilities. Many people like to think of elements of W as possible worlds; during betting it is not known which of them is the actual world. Afterwards reality is uncovered and all the parties involved know of some member A of $At_{\mathcal{F}}$ that it obtains. If $A \subseteq E$, then E is true in the actual world: therefore, a bought bet for E is won and the buyer receives the stake. If $A \not\subseteq E$, then E is false in the actual world; a bought bet for E is lost and the buyer receives nothing.

It is conceivable that for a given betting quotient function a set of bets exists which ensures the agent's loss whatever happens. That is, for any $A \in At_{\mathcal{F}}$, the profits from the stakes of won bets are strictly lower than the losses resulting from the prices paid for all bought bets. Such a set of bets is called a *Dutch Book* and is the topic of the next definition.

Definition 4 (Dutch-book(able)): Let q be a betting quotient function, $A \in At_{\mathcal{F}}$ and $\mathcal{B}$ be a set of bets. q's *profit from* $\mathcal{B}$ *if* A *is true* is defined as

$$\mathrm{Prf}_q(A, \mathcal{B}) := \sum_{E: E \in \mathcal{B}, A \subseteq E} (1 - q_E) S_E + \sum_{E: E \in \mathcal{B}, A \not\subseteq E} -q_E S_E.$$

A set of bets $\mathcal{B}$ is a *Dutch Book* for q if and only if for any $A \in At_{\mathcal{F}}$ $\mathrm{Prf}_q(A, \mathcal{B}) < 0$.

A betting quotient function is called *Dutch-bookable* iff a Dutch Book exists for it.

One popular term for "not Dutch-bookable" in the literature is "coherent". I will also occasionally use it.

The remarkable theorem whose proof and import is the topic of this chapter runs as follows:

Theorem 1 (Dutch Book Theorem). *A betting function* q *is not Dutch-bookable if and only if it is a finitely additive probability function.*

The discovery of the DBT has been attributed to many authors, most frequently to Ramsey (1931, originally published in 1928) and de Finetti; some even write of the "Ramsey-de Finetti theorem". De Finetti definitely proved it in his 1937/1964 paper, which we will discuss in Section 2.1.1.2. I do not see an argument for the DBT in Ramsey's paper, just some suggestions. Not being a historian, I will only say that if indeed Ramsey proved the DBT, he did not manage to make it deservedly well known, since in 1955 Shimony, Lehman and Kemeny published three independently discovered proofs of results which from the modern standpoint and after some work can be seen as corresponding to the DBT in the *Journal of Symbolic Logic*.

We will divide the DBT into two lemmas, corresponding to the "Forward" (left to right; violation of probability axioms implies Dutch-bookability) and "Converse" (right to left; satisfaction of probability axioms gurantees un-Dutch-bookability) directions. It is typical for modern texts on the topic only to talk about the "Forward" direction, since it can be easily illustrated with examples. In comparison, the proof of the "Converse" direction has been "surprisingly neglected" (Hájek (2008), p. 796), therefore I will devote the whole next section to it.

Lemma 1 ("Forward" Dutch Book Theorem). *If a betting function* q *is not Dutch bookable, it is a finitely additive probability function.*

Proof. We will prove the contrapositive: if a betting function q violates any of the axioms (A1)–(A3), a Dutch Book for it exists.

Suppose q violates (A1). Therefore, for some $A \in \mathcal{F}$, $q_A < 0$. Let S be any real number lower than 0. $\{\langle A, S\rangle\}$ is then a Dutch Book for q.

Suppose q satisfies (A1) but violates (A2). Then either $q_\top > 1$ or $0 \leqslant q_\top < 1$. If the former is true, take any $S > 0$. If $0 \leqslant q_\top < 1$, take any $S < 0$. Your choice of S leads to a Dutch Book for q of the form $\{\langle \top, S\rangle\}$.

Suppose q violates (A3), that is, there are two propositions $A, B \in \mathcal{F}$ such that $A \cap B = \emptyset$, but $p(A \cup B) \neq p(A) + p(B)$. If $p(A \cup B) < p(A) + p(B)$, then set S to be any strictly positive real number. If $p(A \cup B) > p(A) + p(B)$, then set S to be any strictly negative real number. Set $S_{A\cup B} := -S$. Your choice of S and $S_{A\cup B}$ leads to a Dutch Book for q of the form $\{\langle A \cup B, S_{A\cup B}\rangle, \langle A, S\rangle, \langle B, S\rangle\}$. □

In the next section I will present three ways of proving the Converse Dutch Book Theorem. (In fact, the presented arguments of Freedman (Section 2.1.1.4) and de Finetti (2.1.1.2) prove both directions of the DBT, but all in due course; the "added value" will be the converse direction.)

2.1.1. Proving the (Converse) Dutch Book Theorem

2.1.1.1. Kemeny: let us clearly employ all the axioms

My presentation of the theorem is based on Kemeny (1955) in terms of the argument. The framework, however, is different. First, Kemeny was concerned with two-argument confirmation functions in the style of Carnap; here the reader will find the argument set in the hopefully familiar context of classical probability functions over fields of sets (Boolean algebras). Second, I will be using the three Kolmogorovian axioms (A1)–(A3) listed above, which are currently considered classical (although in subsection 2.1.1.2, for presentation of the de Finetti argument we will move to a two-axiom setting), while Kemeny uses five axioms. The "additional" ones, not needed once we move to Boolean algebras and classical probability functions, concern conditional probability and probabilities of equivalent propositions. Third, Kemeny uses sets of bets including bets bought both for and against the same event, allowing the stakes to differ between the two cases. I have already mentioned that this is not needed, since anything we need to say using a bought bet against some E we can say using a bought bet for E, and two bets for E with different stakes are equivalent to a single bet for E. Therefore, for any event we can consider just a single bet for it, which simplifies the presentation of the argument.

Note the following trivial fact:

Fact 2. *If a betting function* q *is Dutch-bookable, there exists a Dutch Book for* q *which is a full set of bets.*

Proof. A set of bets $\mathcal{B}$ which is a Dutch Book for q can be extended to a full set of bets $\mathcal{C} := \mathcal{B} \cup \{\langle E, 0\rangle | E \notin \mathcal{F}\}$. Obviously, $\mathcal{C}$ is also a Dutch Book for q. □

By Fact 2, to show that some betting function q is not Dutch-bookable, it is enough to prove that there is no Dutch Book for it by means of a full set of bets.

For a given betting function q and a full set of bets $\mathcal{B}$ with stakes for each proposition E equal to S_E, q's profit if some $A \in At_{\mathcal{F}}$ is true equals, by Definition 4, the following:

$$\mathsf{Prf}_q(A, \mathcal{B}) := \sum_{E:A\subseteq E} (1 - q_E)S_E + \sum_{E:A\nsubseteq E} -q_E S_E. \tag{2.1}$$

Let us define Prf as the expectance of q's profits:

$$\mathsf{Prf} := \sum_{A\in At_{\mathcal{F}}} \mathsf{Prf}_q(A, \mathcal{B}) \cdot q_A. \tag{2.2}$$

In the argument below I have marked in bold the places in which the axioms are invoked for the first time. The crucial lemma will be the following:

Lemma 3. *Suppose a betting quotient function* q *satisfies the probability axioms (A1)-(A3). Then* $\mathsf{Prf} = 0$.

Proof. By elementary transformations,

$$\begin{aligned}
\mathsf{Prf} &= \sum_{A\in At_{\mathcal{F}}} \Big(\sum_{E:A\subseteq E} q_A(1-q_E)S_E + \sum_{E:A\nsubseteq E} q_A(-q_E)S_E \Big) \\
&= \Big(\sum_{A\in At_{\mathcal{F}}} \sum_{E:A\subseteq E} q_A(1-q_E)S_E \Big) + \Big(\sum_{A\in At_{\mathcal{F}}} \sum_{E:A\nsubseteq E} q_A(-q_E)S_E \Big) \\
&= \Big(\sum_{E\in\mathcal{F}} \sum_{A\in At_{\mathcal{F}}, A\subseteq E} q_A(1-q_E)S_E \Big) + \Big(\sum_{E\in\mathcal{F}} \sum_{A\in At_{\mathcal{F}}, A\nsubseteq E} q_A(-q_E)S_E \Big) \\
&= \sum_{E\in\mathcal{F}} \Big(\sum_{A\in At_{\mathcal{F}}, A\subseteq E} q_A(1-q_E)S_E + \sum_{A\in At_{\mathcal{F}}, A\nsubseteq E} q_A(-q_E)S_E \Big) = \\
&= \sum_{E\in\mathcal{F}} \Big(\underbrace{\Big(\sum_{A\in At_{\mathcal{F}}, A\subseteq E} q_A(1-q_E) + \sum_{A\in At_{\mathcal{F}}, A\nsubseteq E} q_A(-q_E) \Big) \cdot S_E}_{=0} \Big).
\end{aligned}$$

We will show that for any $E \in \mathcal{F}$, the underbraced expression is indeed equal to 0. Fix, then, some $E \in \mathcal{F}$. Note (**here we use normalization and**

additivity) that $1 = q_\top = q_{E \cup \neg E} = q_E + q_{\neg E}$, therefore $q_{\neg E} = 1 - q_E$. From additivity we again see that

$$\sum_{A \in At_{\mathcal{F}}, A \subseteq E} q_A = q_E$$

and

$$\sum_{A \in At_{\mathcal{F}}, A \nsubseteq E} q_A = q_{\neg E}.$$

Therefore, the underbraced expression equals $q_E(1 - q_E) - q_E(1 - q_E) = 0$. This holds for any choice of $E \in \mathcal{F}$, therefore $\mathrm{Prf} = 0$. □

Note that only two axioms were used in the proof of the above lemma; nonnegativity was not needed. We will now show the Converse DBT:

Lemma 4 (Converse Dutch Book Theorem). *Suppose a betting quotient function* q *satisfies the probability axioms (A1)–(A3). Then there is no Dutch Book for* q.

Proof. For a *reductio*, assume that q satisfies the probability axioms, and yet there is a Dutch Book for q. Then there exists a full set of bets $\mathcal{B}$ such that $\forall A \in At_{\mathcal{F}}\ \mathrm{Prf}_q(A, \mathcal{B}) < 0$. We know (from normalization and additivity) that $1 = q_\top = q_{\cup\{A\}_{A \in At_{\mathcal{F}}}} = \sum_{A \in At_{\mathcal{F}}} q_A$, therefore, for some $A \in At_{\mathcal{F}}\ q_A > 0$. This, coupled with the assumption that $\forall A \in At_{\mathcal{F}}$ $\mathrm{Prf}_q(A, \mathcal{B}) < 0$ and the fact that **from nonnegativity** we know that $\forall A \in At_{\mathcal{F}}\ q_A \geqslant 0$, allows us to infer that $\mathrm{Prf} < 0$, which contradicts Lemma 3. □

The proofs of Lemmas 1 and 3 together constitute proof of the Theorem 1, that is, the Dutch Book Theorem.

In the next section I will present a reconstruction of de Finetti's argument with a beautiful application of linear algebra. Next, we will see an argument inspired by Freedman (2003) in which the notion of expected value occupies the spotlight.

2.1.1.2. De Finetti: linear algebra

In this subsection I will present an argument for probabilism which can be recovered from de Finetti (1937/1964) (it is the one alluded to by Gillies (2000) in the passage I mentioned in the last chapter). Notice first that the notion of a finitely additive probability measure can be equivalently formulated using only two axioms.[2] A *partition* of the Boolean algebra $\mathcal{F}$ is a family of sets $\mathcal{E} \subseteq \mathcal{F}$ which are pairwise disjoint and exhaust the whole $\mathcal{W}$, that is, (1) $\forall E, F \in \mathcal{E}$, if $E \neq F$ then $E \cap F = \emptyset$, and (2) $\cup\mathcal{E} = \top$. (If $\mathcal{F} = \mathcal{P}(W)$, we will use the terms "partition of $\mathcal{F}$ and "partition of W" interchangeably.)

Fact 5. *A function* $p : \mathcal{F} \to \mathbb{R}$ *is a finitely additive probability function iff the following two axioms are satisfied:*

(B1) for any $A \in \mathcal{F}$, $p(A) \geqslant 0$;

(B2) for any partition $E = \{E_1, \ldots, E_n\}$, $p(E_1) + \ldots + p(E_n) = 1$.

Proof. (B1) is identical to (A1). (A3) implies that for any partition $E = \{E_1, \ldots, E_n\}$, $p(E_1) + \ldots + p(E_n) = p(\top)$, which, by (A2), equals 1: therefore (A2) and (A3) together imply (B2).

Consider any $A, B \in \mathcal{F}$ such that $A \cap B = \emptyset$. Since $\{A \cup B, \neg(A \cup B)\}$ and $\{A, B, \neg(A \cup B)\}$ are partitions, from (B2) we know that $p(A \cup B) + p(\neg(A \cup B)) = 1 = p(A) + p(B) + p(\neg(A \cup B))$; therefore $p(A \cup B) = p(A) + p(B)$ and so we have established that (B2) implies (A3). Since $p(\top) = p(A \cup \neg A)$, by (A3) $p(\top) = p(A) + p(\neg A)$, which by (B2) equals 1. Therefore (B2) implies (A2). □

What follows is a proof of the Converse Dutch Book Theorem based on de Finetti's insights from de Finetti (1937/1964). I now need to clarify my statement above that "de Finetti definitely proved" the DBT in that paper. This statement is warranted, in my opinion, since de Finetti offers an observation from the realm of linear algebra which can be used as the main engine behind the proof of the Converse DBT; while he writes that his result shows a necessary and sufficient condition for coherence, it does not actually do it when read completely literally.

---

[2] For more reading about the subject of axiomatizing elementary probability theory I'd recommend Goosens (1979).

First, it shows that a quotient function satisfying (B1) and (B2) is not Dutch-bookable using a set of bets on elements of some partition, but *a priori* different sets of bets should also be considered. Second, it shows that violation of (B2) leads to a Dutch Book, but is silent regarding the violation of (B1). I suppose de Finetti considered the latter to be trivial.

Theorem 2 (Converse Dutch Book Theorem, de Finetti style). *Suppose a betting quotient function* q *satisfies the probability axioms (B1) and (B2). Then there is no Dutch Book for* q.

Proof. Since q satisfies (B2), as already mentioned, q is finitely additive. Therefore, any bet on $E = \bigcup\{A_i\}_{i\in\{1,\ldots,k\}}$ ($A_i \in At_{\mathcal{F}}$ for any i) with a stake S is equivalent payoff-wise to the set of k bets $\{\langle A_i, S/k\rangle\}_{i\in\{1,\ldots,k\}}$ (since $q_E = q_{A_1} + \ldots + q_{A_k}$). Therefore, any Dutch Book for q is equivalent payoff-wise to a Dutch Book consisting solely of bets made for all elements of $At_{\mathcal{F}}$ (some of these can have null stakes).

Suppose $\mathcal{F}$ has n atoms. For convenience let us write q_i for q_{A_i}. Let S_i be the stake of the bet for A_i. q's profit if A_h is true equals

$$G_h = S_h - \sum_{i=1}^{n} q_i S_i.$$

Consider the system of n equations of this form for $h \in \{1,\ldots,n\}$ and treat the n stakes as the unknowns. Suppose, given a betting quotient function q, you have a specific set of gains in mind (for example, all negative, i.e., a Dutch Book); is it possible to find a set of stakes ensuring precisely *these* gains? That is, does the aforementioned system of equations have a solution?

Note the determinant:

$$\begin{vmatrix} 1-q_1 & -q_2 & \cdots & -q_n \\ -q_1 & 1-q_2 & \cdots & -q_n \\ \cdots & \cdots & \cdots & \cdots \\ -q_1 & -q_2 & \cdots & 1-q_n \end{vmatrix} = 1-(q_1+\ldots+q_n).^3$$

(Note that if (B2) is violated, then that determinant is not equal to zero, that is, a solution exists: it is possible to set the stakes to achieve

[3] If you would like to have a hint as to how to compute it, this is Exercise 6.2.14 in Meyer (2000), p. 484.

arbitrary gains, and so q is Dutch-bookable. This is a part of the proof of the Forward Dutch Book Theorem which de Finetti notes.)

Since we have assumed that q satisfies (B2), we can use the fact that $\sum_{i=1}^{n} q_i = 1$ to show that

$$\sum_{h=1}^{n} q_h G_h = \Big(\sum_{h=1}^{n} q_h S_h\Big) - \Big(\sum_{h=1}^{n} q_h \sum_{i=1}^{n} q_i S_i\Big) = 0,$$

from which it follows by (B1) that not every G_h can be negative. Therefore, no Dutch Book for q exists. □

De Finetti only points out that the determinant for a system of equations corresponding to gains from bets for all elements of some partition $E_1, \ldots, E_k$ equals $1 - (q_1 + \ldots + q_k)$. This immediately gives us a part of the Forward Dutch Book Theorem, that is, not being Dutch-bookable requires (B2): if (B2) is violated, then there exists a partition for which it is violated, and we can construct a Dutch Book using it. However, this, even coupled with (B1), does not give us the Converse Dutch Book Theorem: we only see that q cannot be Dutch-booked using a set of bets consisting exclusively of bets for elements of some partition. One way to obtain complete proof is to point out, as I have done, that in the context of (B2), which implies additivity, any Dutch Book would be equivalent to one concerning only the elements of the maximally fine-grained partition. De Finetti does not mention this; again, he was possibly thinking this was too trivial to address.

2.1.1.3. Digression: de Finetti's determinant argument for conditionalization

I would like to point out another application of linear algebra in de Finetti (1937/1964) which might be of interest for different reasons than de Finetti himself proposed. His proclaimed goal (p. 68) is—assuming the rationality of the probability axioms discussed so far has already been established—to defend the choice of the "axiom" governing conditional probability, namely, that the conditional probability $p(E'|E'')$ is to be equal to $p(E' \wedge E'')/p(E'')$; we will speak about betting quotients instead (for de Finetti they seem to have been the same thing). The idea is to think about $E'|E''$ as a "tri-event" and a bet for it as a conditional bet which is won if $E'' \wedge E'$, lost if $E'' \wedge \neg E'$, and void if $\neg E''$. Suppose the betting quotient for $E'|E''$ is q. This leads to the following payoff table for a bet for $E'|E''$ with a stake S:

BOUGHT	$E'' \wedge E'$	$E'' \wedge \neg E'$	$\neg E''$
bet for $E'\|E''$	$(1-q)S$	$-qS$	0

Suppose, then, that $E' \subseteq E''$ (this does not diminish the generality of the argument), that q' is the betting quotient for E' and q'' for E''. De Finetti shows that coherence requires that $q' = q \cdot q''$. Consider three bets: for E' with the stake S', for E'' with the stake S'', and for $E'|E''$ with the stake S. The three possible states of affairs of interest to us lead to the following payoffs:

$$E' : \; G_1 = (1-q')S' + (1-q'')S'' + (1-q)S$$

$$E'' \wedge \neg E' : \; G_2 = -q'S' + (1-q'')S'' - qS$$

$$\neg E'' : \; G_3 = -q'S' - q''S''$$

Treating this as a system of equations with the stakes as unknowns, calculate the determinant:

$$\begin{vmatrix} 1-q' & 1-q'' & 1-q \\ -q' & 1-q'' & -q \\ -q' & -q'' & 0 \end{vmatrix} = -q' \begin{vmatrix} 1-q'' & 1-q \\ 1-q'' & -q \end{vmatrix} + q'' \begin{vmatrix} 1-q' & 1-q \\ -q' & -q \end{vmatrix} =$$

$$= q' - q \cdot q''.$$

So, if $q' \neq q \cdot q''$, then arbitrary gains, for example, exclusively negative gains, can be obtained, and so the betting quotient function is Dutch-bookable.

De Finetti seems to consider this to be justification for a particular, now standard, treatment of conditional probability. But I would like to propose that, in addition to this, the linear algebra insight can be used in an argument for a diachronic requirement of rationality; my discussion needs to remain sketchy here because I will not propose a formal definition of a diachronic Dutch Book which would deal with betting quotient functions of an agent at different times, hoping that the idea will be intuitive enough. The requirement just mentioned is that of conditionalization: upon learning that E, an agent should revise his betting quotient in a proposition A from q_A to $q_{A \wedge E}/q_E$. (We will encounter the rule of conditionalization in Chapter 5.) This is because otherwise, if the revised betting quotient is some $q \neq q_{A \wedge E}/q_E$, then we could apply the above argument putting $q'' := q_E$ and $q' := q_{A \wedge E}$ and find a set of stakes leading to a set of bets such that after it is decided whether E actually takes place and then if A is true, the agent's profits are negative regardless of which of the options materializes.

It is typical in the literature to cite Teller (1973) as first reporting a Dutch Book argument to the effect that violating conditionalization entails Dutch-bookability, and to attribute the original argument to Lewis (his note regarding this was eventually published as Lewis (1999)). However, it seems to me that we can use de Finetti's insights as the basis for a more concise argument. Whether it also gives us the converse result that adhering to Conditionalization makes one un-Dutchbookable is something I would like to study in the future.

2.1.1.4. Freedman: just use the expected value

I trust that at this point the reader has seen enough of the arguments for the DBT to be convinced that it holds. I have tried to indicate carefully where exactly the axioms are called upon in the arguments. I will now present reasoning based on the argument by Freedman (2003)[4]; the original author mentions no axioms and relies on the notion of expected value (see below) to do the job. Since Freedman uses odds instead of betting quotients and requires them to be positive and finite, propositions for which a betting quotient function obtains a value of 0 or 1 (which might be propositions different from $\emptyset$ or $\top$!) are outside the scope of his reasoning. I will generalize his result in that respect and fill in all the gaps so that it is clearly seen where the axioms are employed. Since I have already given two proofs of the Converse DBT, I have opted for a slightly less formal style in this subsection.

---

[4] Who claims to "sketch the mathematics behind de Finetti's argument for the Bayesian position".

A *random variable* α is simply a function $W \to \mathbb{R}$; it attaches a number to each possible world. For example, if a betting quotient function is fixed, then any bet is a random variable—its value is the profit depending on whether the given proposition is true in the given world. The *range* of α, $\mathrm{ran}(\alpha)$, is the set of values obtained by α. (For those with a mathematical background, recall that we are dealing with finite structures only, so I'm skipping the requirements regarding the measurability of the values of α^{-1}, etc.)

What the random variable is is independent of any probability function; it can be thought of as coding the functional dependence of some value on the state of the world. Therefore, instead of defining a random variable in the context of a particular probability space (as done e.g. in Rosenthal (2006)) I will proceed like e.g. Schervish and DeGroot (2014) and define it as a function with a domain which might be a sample space of various probability spaces.

Definition 5 (Random variable): Let W be a finite nonempty set. A *random variable on* W is a function $\alpha : W \to \mathbb{R}$. The *range* of α is defined as

$$\mathrm{ran}(\alpha) = \{a \in \mathbb{R} \mid \exists w \in W \alpha(w) = a\}.$$

Suppose $\langle W, \mathcal{F}, p\rangle$ is a finite probability space and α is a random variable on W. The *expected value of* α *according to* p is defined as

$$EX_p(\alpha) = \sum_{a \in \mathrm{ran}(\alpha)} p(\alpha^{-1}(a)) \cdot a.$$

Note that for any random variable α, the set $\{\alpha^{-1}(a) | a \in \mathrm{ran}(\alpha)\}$ forms a partition of W.

One and the same random variable may have different expected values according to different probability functions. For example, if W is the six-element set of all possible outcomes of a roll of a six-sided die, and the random variable attaches the value 1 to each of the worlds with an even outcome and 0 to all other worlds, the variable has an expected value of 1/2 if the die is fair. However, if the die is loaded, and some outcomes have different probabilities of occurring, then the exact same random variable might have a different expected value.

Given a betting function q, it may be natural to think of $\sum_{a \in \mathrm{ran}(\alpha)} q_{\alpha^{-1}(a)} \cdot a$ as something like "the expected value of α according to q"; furthermore, if that q codes some agent's approach to bets, and α corresponds to a bet, it may be intuitive that the formula gives the expected value of the bet according to that particular agent. We will eventually settle for this way of thinking, as is actually typical in the literature. When discussing the Dutch Book Argument (as opposed to the Theorem) below, we will need to talk about the expected values of bets according to various agents. However, we need to be conscious of the fact that when doing so we will be departing from the usual mathematical notion of expected value, which requires the function whose values are multiplied by the values of the variable to be a probability function. We will return to these issues repeatedly below.

A sum of two random variables obviously is a random variable; its expected value is the sum of the expected values of the two. Assume again that some betting function q is fixed. Since any single bet is a random variable, then a finite set of bets (and we only consider these) is also one. If that set of bets is a Dutch Book for q, it means that as a function of W it has exclusively negative values. Therefore, it has a negative expected value according to *any* probability function. Therefore, to prove the Converse DBT, that is, to show that a probabilistic q is not Dutch-bookable, it is enough to show that for any set of bets there exists a probability function according to which this set has a nonnegative expected value. It will turn out that the probability function doing the trick (that is, leading to a nonnegative expected value of any set of bets) is q itself.

We will use the notion of the *indicator function* for a proposition $E \in \mathcal{F}$, which is the function $\mathbf{1}_E : W \to \{0, 1\}$ such that

$$\mathbf{1}_E(w) = \begin{cases} 1 & \text{if } w \in E \\ 0 & \text{otherwise.} \end{cases}$$

Assume then that q is a betting quotient function of some agent and is also a probability function. We will show that q is not Dutch-bookable.

Consider the following random variable ϕ_A, which attaches to each $w \in W$ the profit from betting for E with the stake S:

$$\phi_A := \mathbf{1_A}(1 - q_A)S - (1 - \mathbf{1_A})q_A S.$$

The total payoff function for any full set of bets is another random variable on W, equal to

$$\Theta := \sum_{A \in \mathcal{F}} \phi_A.$$

Let us calculate the expected value of ϕ_A according to q:

$$EX_q(\phi_A) = q_A(1 - q_A)S + q_{\neg A}(-q_A)S) = q_A(1 - q_A)S - (1 - q_A)q_A S = 0,$$

where we have, as above in presenting Kemeny's argument, used **normalization** and **additivity** to substitute $1 - q_A$ for $q_{\neg A}$.

Since Θ is a finite sum of random variables with expected value 0, $EX_q(\Theta) = 0$. The values of Θ are the possible profits of the agent from the given set of bets. From $EX_q(\Theta) = 0$ and **nonnegativity** we see that these values cannot all be negative, therefore q is not Dutch-bookable.

It can be protested that in the argument just given the role of the probability axioms is somewhat unclear, since we rely on q being a probability function already when starting to calculate the expected value of ϕ_A with respect to q. We can re-run the argument, keeping the old definitions of the "profit from bet for A" random variable ϕ_A and "total profit from the given set of bets" random variable Θ, but using the following notion, which poses no similar restrictions:

Definition 6 (Generalized expected value): Let W be a finite nonempty set, $\mathcal{F} = \mathcal{P}(W)$ and let α be a random variable on W. Let r be any function from $\mathcal{F}$ to $\mathbb{R}$. The *generalized expected value of* α *according to* r is defined as

$$GEX_r(\alpha) = \sum_{a \in \mathrm{ran}(\alpha)} r(\alpha^{-1}(a)) \cdot a.$$

One (instructively mistaken) train of thought could be the following: It would seem that, like before, we can arrive at $GEX_r(\Theta) = 0$ just by requiring from r that for any $A \in \mathcal{F}$ $r(A) + r(\neg A) = 1$, the condition of "negation coherence", which will be of great interest to us in the next

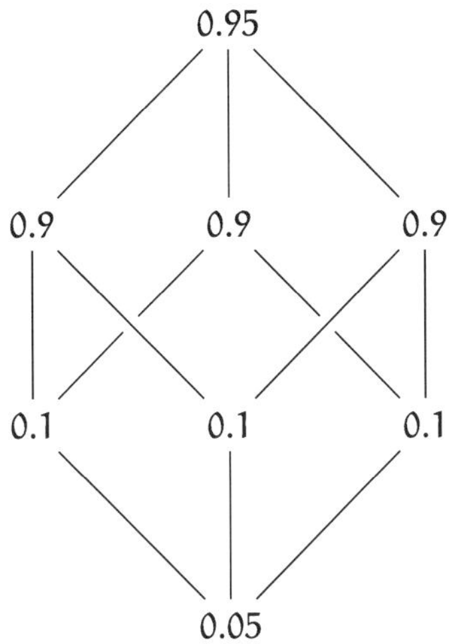

Figure 2.1. A function r which satisfies nonnegativity and negation coherence, but violates additivity

	α	β	$\alpha+\beta$
w_1	1	5	6
w_2	4	2	6
w_3	0	4	4

Table 2.1. Singletons of the w_i's are atoms of the algebra depicted in Figure 6.1. This table defines random variables α and β for which $\mathrm{GEX}_r(\alpha)+\mathrm{GEX}_r(\beta) = 0.5+1.1 = 1.6 \neq 5.8 = 5.4+0.4 = \mathrm{GEX}_r(\alpha+\beta)$

section. Then we could add the requirement of nonnegativity of r to be able to infer from the fact that for any set of bets $\mathrm{GEX}_r(\Theta) = 0$ the conclusion that r is not Dutch-bookable. We would therefore seemingly establish the Converse DBT without appealing to additivity.

This, however, is a mistake, since the generalized expected value of a sum of two random variables does not in general equal the sum of the generalized expected values of these variables. Figure 2.1 and Table 2.1 depict an example using a function r which satisfies both nonnegativity and negation coherence, but violates additivity. To move from $\mathrm{GEX}_r(\phi_A) = 0$ to $\mathrm{GEX}_r(\Theta) = 0$, we can appeal to additivity; let us note the following elementary fact:

Fact 6. *Let W be a finite nonempty set, $\mathcal{F} = \mathcal{P}(W)$ and let α and β be random variables on W. Let r be any function from $\mathcal{F}$ to $\mathbb{R}$. Then if r is additive, $GEX_r(\alpha) + GEX_r(\beta) = GEX_r(\alpha + \beta)$.*

Therefore, I propose one "conceptually clean" form of the argument for the Converse DBT which appeals to the notion of the expected value being as follows:

1. use the notion of General Expected Value instead of Expected Value;
2. appeal to negation coherence to argue that $\mathrm{GEX}_r(\phi_A) = 0$ for any A;
3. appeal to **additivity** to obtain from this the conclusion that $\mathrm{GEX}_r(\Theta) = 0$;
4. appeal to **nonnegativity** to conclude that r is not Dutch-bookable.

Since additivity and negation coherence are together equivalent to additivity and **normalization**, this is another way of establishing the Converse DBT.

*

We have seen three ways of establishing the Dutch Book Theorem. I will now argue that the Dutch Book *Argument* fails and offer a repaired, weaker argument.

2.2. The Dutch Book Argument and why it fails

The Dutch Book Argument tries to establish the norm of Probabilism, which says that the degrees of belief (or, interchangeably, "credences") of a rational agent should conform to the classical probability axioms. As already mentioned, the DBT is at the heart of the argument. But something needs to be said about the relationship between credences and betting quotients. I will argue now that there is a largely unexplored lacuna here which leads to the demise of the Dutch Book Argument for Probabilism in its full generality; at the very least, while us probabilists can use it to reinforce our beliefs that we're doing the right thing (but bordering on a *petitio principii*), it cannot in general be used to convince a non-probabilist.

Before we continue, let us define the concept of an agent's belief space and degree of belief function.

Definition 7 (Belief space): A belief space is a tuple $\langle W, \mathcal{F}, \mathrm{b}\rangle$, where W is a nonempty finite set, $\mathcal{F}$ is the power set of W (containing 'propositions'), and b is a function from $\mathcal{F}$ to $\mathbb{R}$, called the *degree of belief function*.

A degree of belief function may also be called a "belief function" or a "credence function". Mathematically, of course, it is indistinguishable from a betting quotient function. However, the relation of one to the other in the case of a given agent is not immediate.

The problem can be seen already in the case which was the easier part of the DBT: showing the consequences of violating the probability axioms. Typically, the Dutch Book Argument will attempt to proceed as follows:

1. Assume that the agent's degree belief function violates the probability axioms;
2. define the agent's betting quotient for a proposition E to equal one such that the agent expects neither profit or loss (such a bet is called "fair" according to the agent);
3. observe that this leads to the identification of the agent's credences with his betting quotients;
4. appeal to the DBT to point out that the agent faces sure loss;
5. conclude that the agent's degrees of belief are irrational.

Step 2 can be skipped by those who wish to identify degrees of belief with betting quotients from the start. I fail to see what this form of strict operationalism gives us apart from a pleasing but fleeting illusion of empiricism. I suggest that, in the spirit of Eriksson and Hájek (2007), we try to keep an open mind regarding what degrees of belief *are*, and investigate the relationship between them and betting quotients using just the basic assumption that *whatever* they are, they can be expressed by a real number.

To hopefully elucidate a possibly disturbing aspect of step 2 above, it is quite common to furnish a bridge between credences and betting quotients by means of "fairness"; for example, suppose your $b(A) = 1/3$ and your $b(\neg A) = 2/3$. Then if you consider a betting quotient $q_A = 1/2$, you expect the bet for A with a stake equal to 1 to end with your "profit" of $1/3 \cdot (1 - 1/2) + 2/3 \cdot (-1/2) = -1/6$, which is not "fair" according to the way the term is used in a big portion of Dutch-book related literature (and not, of course, in real life). The choice of $q_A = 1/10$ will lead to positive expected profit, which is also not "fair". For a different angle, notice that it is not the quotient you would choose if you knew you would take part in the bet, but did not know whether you would buy or sell, and would like to avoid loss. The one and only betting quotient satisfying this requirement equals $b(A)$.

The DBA has been criticized in many ways which I will not attempt to rehearse here; instead I will try to promote my own criticism. I suggest the interested reader go through the concise treatment given in Chapter 3.4 of Childers (2013) and explore Joyce (1998) as the source of the modern discussion about the supposedly pragmatic nature of the argument being a defect from the point of view of epistemology.

Imagine, then, an attempt to convince your nonprobabilist friend of the error of his ways. Your friend claims that his degree of belief function b is not a probability function and that is how he wants it to be. You go through all the steps of the above argument to show that he is vulnerable to a Dutch Book, a sure-loss betting setup. You discover, though, that he is not convinced, and his reasons are none of the usual criticisms of the DBA from the literature. He agrees that his betting quotient for the bet for A with a stake S is that particular q for which $b(A) \cdot (1 - q)S + b(\neg A) \cdot (-q)S$ is equal to 0, and he agrees with the interpretation that it is the quotient for which he expects the profit from the bet to equal 0. Still, he claims that his betting quotients are different from his degrees of belief.

He claims that this is because in his case $b(A) + b(\neg A)$ is in general not equal to 1, but rather, that for each proposition A there is a non-zero number r_A for which it holds that $b(A) + b(\neg A) = r_A$; of course some of these numbers may be equal to 1. Therefore, his betting quotient for the bet for a proposition A is in general equal to $^{b(A)}/_{r_A}$. In other words, the agent claims he is "negation incoherent", that he violates the condition we have already met in the previous Section (first discussed in print by Hedden (2013)). Can we re-run our argument using betting quotients of an agent who is negation incoherent, and so his degree of belief function violates at least one of the conditions of normalization and additivity? We will see below that in *some* cases we can: it is possible to rigorously specify the class of nonprobabilistic agents susceptible to the argument.

In fact, I find it quite puzzling that the prevailing consensus in philosophy has been that violation of probability axioms by one's belief function is a sign of one's irrationality. Take the normalization axiom. Surely the choice of number 1 as the probability of tautologies (or "certain events", etc.) is conventional; there are plenty of numbers which would do similarly well. (Some calculations might become more cumbersome for us trained in the "classical ways", but can we say with full confidence that classical probability axioms offer us the simplest way of reasoning about these matters?) But apparently, if we set our credences in tautologies to something different from 1, we can be Dutch-booked. For me this is one of the main reasons to suspect the Dutch Book Argument.

To come back to negation incoherence, in my opinion it can be seen as another reason for which "betting odds and credences come apart" (Bradley & Leitgeb (2006); Rees (2010)). Unlike the reasons from the cited works, though, it has nothing to do with the issues related to self-location (see Elga (2000) for one of the main sources of modern discussions of that subject). As the reader hopefully is convinced by now, there is nothing wrong with the classical Dutch Book *Theorem*; however, it concerns betting quotients, while the Dutch Book *argument* tries to reach some conclusion about credences. If there are situations in which these two "come apart", then we should try to describe a rigorous link between them and reassess the strength of the argument.

The idea is, then, that an agent has some belief function, but whether he is Dutch-bookable or not depends on his betting quotient function.

The link between the two comes in the form of "being induced" in the form of the following definition:

Definition 8 (Induced betting quotient, Dutch-bookable belief function): A belief space $\langle W, \mathcal{F}, b \rangle$ induces a betting quotient $q : \mathcal{F} \to \mathbb{R}$ if for any $A \in \mathcal{F}$:

1. $b(A) + b(\neg A) \neq 0$,
2. $q(A) = \frac{b(A)}{b(A)+b(\neg A)}$.

A belief function (or space) is called *Dutch-bookable* if and only if its induced betting quotient function is Dutch-bookable as per Definition 4.

In light of this definition we can, I think, reasonably say that $q(A)$ is the betting quotient which makes a bet for or against A such that an agent with a belief function b expects it to have value 0. More formally, it is the quotient for which the bet has Generalized Expected Value 0 from the perspective of b (see Definition 6). It also follows that if a belief space induces a betting quotient function, that is, if the first condition of the above definition holds, then that function is unique. So, if a belief function b is such that for some A $b(A) + b(\neg A) = 0$, it does not induce a betting quotient function at all; otherwise, it induces one and only one.

The question which now arises is the following: are there any non-probabilistic epistemic agents who are not Dutch-bookable? Formally: are there nonprobabilistic belief functions which are not Dutch-bookable by Definition 8? In other words, are there belief functions which violate at least one of the three probability axioms and which induce a not Dutch-bookable betting quotient function, that is, they induce a betting quotient function which is a probability measure? The answer is given by the following theorem:

Theorem 3 (Wroński & Godziszewski (2017)). *The betting quotient function q induced by a belief space $\langle W, \mathcal{F}, b \rangle$ is a classical probability function iff the following conditions hold:*

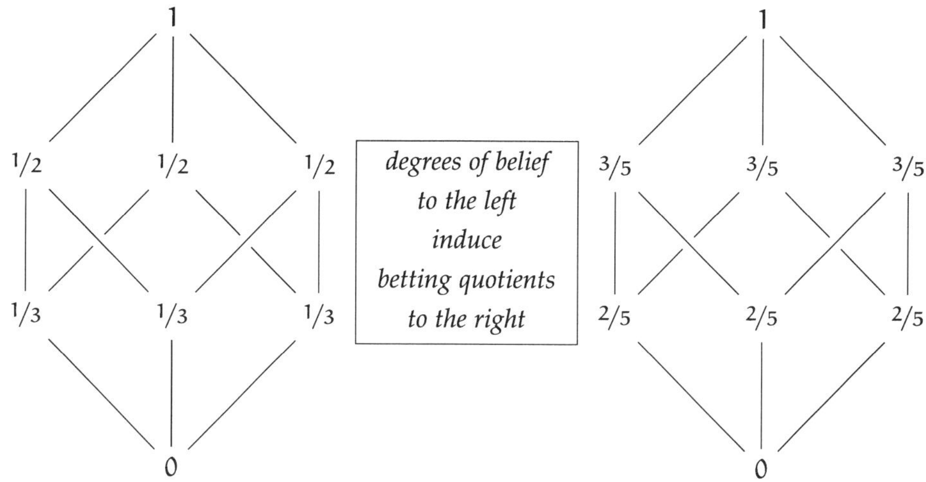

Figure 2.2. A nonprobabilist, Dutch-bookable belief space and its induced betting quotient function. (This example first appeared in Wroński & Godziszewski (2017), but was incorrectly captioned.)

1. $b(\emptyset) = 0$,
2. *for any* A *in* $\mathcal{F}$ $b(A) \cdot b(\neg A) \geqslant 0$,
3. *for any* A *and* B *in* $\mathcal{F}$ *with an empty intersection:*

$$\frac{b(A \cup B)}{b(A \cup B) + b(\neg(A \cup B))} = \frac{b(A)}{b(A) + b(\neg A)} + \frac{b(B)}{b(B) + b(\neg B)}.$$

To see an example of a Dutch-bookable, nonprobabilist belief space, consider the space with three atomic propositions depicted in Figure 2.2, where the left algebra depicts the function b defined on $\mathcal{P}(\{w_1, w_2, w_3\})$, and the right one represents the induced betting quotient function q.[5]

We can see that the betting quotient function q is not additive, and so is not a probability measure, therefore the belief function b which induces it is Dutch-bookable; as the reader may check, it does not satisfy the third condition of Theorem 3.

[5] This and the following example appeared first in Wroński & Godziszewski (2017) and are cited here with corrections.

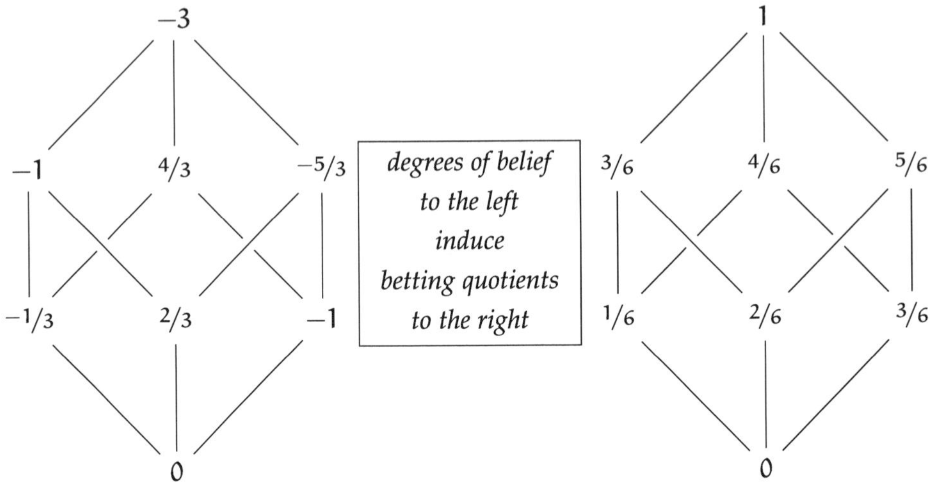

Figure 2.3. A nonprobabilist, un-Dutch-bookable belief space with a "wild" belief function. (This example first appeared in Wroński & Godziszewski (2017).)

However, a nonprobabilistic belief function may be intuitively "wild" and yet induce an un-Dutch-bookable betting quotient function, as evidenced by Figure 2.3.

I should stress here that I do not suggest that possessing negative credences, whatever this could mean, is rationally permissible. My goal here is to establish the exact power of the Dutch Book Argument on the assumption that degrees of belief—whatever they *really* are—are expressible by real numbers, and the agent's betting quotients are induced in the above way, so that "individual bets involved in making the book are fair, which is to say they have an expected value of zero, when calculated using the agent's betting quotients" (Vineberg (2016)). Negative betting quotients, as we have seen, lead to a sure loss. Negative credences might also make the agent exploitable, but this would depend on the particular interpretation of credences one would offer; negative credences by themselves are not enough to render one susceptible to a Dutch Book.

There is something we can say about negative credences, though, which might be seen as somewhat undermining my work in this section, or at least, if we're being positive, rules out one interpretation of negative credences which might initially be seen as intuitively plausible.

Consider, then, the following way of thinking about credences: positive ones are degrees of belief, and negative ones are degrees of *dis*belief. Suppose your $b(A) = 1/4$ and $b(\neg A) = 2/3$. Your induced betting quotient for A equals $q_A = 3/11$; the fact that it is less than $1/2$ is somewhat intuitive since you believe A to a lower degree than $\neg A$. But note that you would obtain exactly the same betting quotient were your credences equal to $b(A) = -1/4$ and $b(\neg A) = -2/3$. This seems counterintuitive to me, since on the assumed interpretation of credences in such a case you *dis*believe A less than $\neg A$, and so according to you a fair price for a bet for A should be higher than half of the stake. Therefore, negative credences are not degrees of disbelief, and if you'd like to furnish an interpretation of the notion (I do not), you should look elsewhere.

2.3. On two notions of expected value used in the literature

I have argued that the DBA is not a valid argument against all non-probabilists. In this section I would like to point out that different approaches to calculating expected value are used in different parts of the literature, and that, at least at first glance, a nonprobabilist might try to use it to defend his or her stance. For example, both the widely cited (Greaves and Wallace (2006), p. 615) and (Pettigrew (2016), Chapter 14) use effectively the following notion of expected value when calculating "expected epistemic utility" (the name of the notion is mine):

Definition 9 (Generalized atomic expected value): Let W be a finite nonempty set, $\mathcal{F} = \mathcal{P}(W)$, and let α be a random variable on W. Let r be any function from $\mathcal{F}$ to $\mathbb{R}$. The *generalized atomic expected value of* α *according to* r is defined as

$$\mathrm{GAEX}_r(\alpha) = \sum_{a \in \mathrm{ran}(\alpha)} \sum_{w \in r(\alpha^{-1}(a))} r(\{w\}) \cdot a.$$

That is, you go over the possible values of the random variable; for each value you consider the subset of W for the elements of which this particular value is obtained; for each element w of such a set you multiply the value of $r(\{w\})$ by $\alpha(w)$: these are the elements of the sum you are calculating.

For example, consider a bet for A with a stake S. It is a two-valued random variable with the generalized atomic expected value equal to

$$\sum_{w \in A} b(\{w\}) \cdot (1 - qS) + \sum_{w \notin A} b(\{w\}) \cdot (-qS), \tag{2.3}$$

while its generalized expected value, as used up to this point, equals

$$b(A) \cdot (1 - qS) + b(\neg A) \cdot (-qS). \tag{2.4}$$

If b is a probability measure, it doesn't matter whether we calculate its expected value, or generalized expected value, or generalized atomic expected value. Therefore, I do not want to impute any error on part of the cited authors since they are using the "atomic" notion when discussing some variants of the conditionalization norm at a point at which probabilism can be already assumed to hold.[6] However, I would like to suggest that employing this way of calculating the expected value is *only* valid in such contexts.

If b is nonadditive, expressions (2.3) and (2.4) might have different values. Accepting GAEX as a legitimate way of calculating expected value potentially opens new avenues for discussion.

For example, a nonprobabilist trying to defend his belief function might point out that when calculating his betting quotient, he "expects" the bet to have the same value as according to GAEX. Therefore, to assess whether he is Dutch-bookable or not we would need to perform different calculations than the ones offered up to now.

However, it is unclear whether this option is really open to the agent. It might be tempting to ask him how exactly he proceeds to assess the quotient for a bet for A. If he claims, for example, that he takes into account how probable he thinks A and $\neg A$ are— which is to

[6] The difference between the two approaches to calculating expected value discussed here is also noted in section 3.3 of Leitgeb & Pettigrew (2010a).

say that he considers his credence in winning the bet and his credence in losing the bet, and his profit in those two cases—it seems intuitive that in calculating the expected value he should use equation (2.4). He might say, though, that he goes through all the possible options, that is, elements of W, and considers his credences in that *that* particular option is the actual world (the $\mathtt{b}(\{w\})$'s) and his profits in such cases, which might be seen to lead to the equation (2.3). This, however, makes all credences in nonatomic propositions irrelevant for betting quotients, even if the bet under consideration concerns such a proposition, which seems to me to be changing the topic: the agent effectively considers only a part of his belief function, while he might be irrational due to the part he leaves out.

To sum up this short section, while I wanted to point out that a different, "atomic" notion of expected value is fruitfully used in other areas of formal epistemology, I would like to stress that it should not be used in discussions regarding probabilism.

*

In this chapter I have tried to convince the reader that the Dutch Book Theorem holds (not that anyone needed convincing—I wanted to elaborate on the formal details), why the classical Dutch Book Argument fails, and how it can be repaired. My arguments regarding the latter are surely unacceptable to anyone who claims that degrees of belief simply *are* betting quotients; however, as already mentioned, I see no benefits to this position. I claim that I have various degrees of belief in many propositions I have never bet on, will never bet on, and could never bet on, all reasonable combinations of modalities of time and real possibility considered. I would therefore like to make a minimal assumption that whatever they really are, they can be expressed by a real number, and see whether I can discover what restrictions on them are imposed by the Dutch Book Argument.

It could be said that Dutch Book considerations should concern ideal agents. In such a case I would have to say that if an ideal agent bets on *everything*, I wouldn't want to aspire to such an ideal (what a waste!), and if such an agent actually does not, but *can* bet on everything, this seems to me to be an aspect of his that is separated from any epistemological considerations.

Chapter 3

The Principal Principle, Best Systems and Humean Supervenience

In this short chapter I will explore a few issues related to the well-known Principal Principle, introduced by Lewis (1986a) as a starting point for analysis of objective chance from a subjective standpoint, and later seen as essential to the problematic task of fleshing out his philosophical program of Humean Supervenience so that it offers an analysis of chance. The Principle connects "reasonable" initial credence functions to knowledge about chances ("The Principal Principle requires that the chancemaking pattern (...) would, if known, correspondingly constrain rational credence" (Lewis (1994))) and so, indirectly, connects such credences with chances themselves. I would like to stress here that Lewis's goal was to elucidate the concept of chance; he claimed the Principle captures "all we know about chance" (p. 87)[1] and wished to use it to learn more about chance. That is, the goal was not—at least not *primarily*—to introduce another item to the list of norms of rational belief discussed in the literature, but rather to point to a principle which might offer us a firmer grip on the elusive concept of (objective) chance.

The Principle has amassed an enormous body of literature I will not attempt to summarize here. (Probably the most extensive study, as

[1] In this chapter all sourceless page references are to Lewis (1986a).

well as a veritable goldmine of creative ideas on the topic, is Masterton (2010).) Its introduction in Lewis (1986a) is a paragon of slow and clear philosophical argumentation; any attempts on my side to sum it up certainly wouldn't be able to hold a candle to it—the paper is huge for a reason. I submit, therefore, that the current chapter is not as self-contained as the previous one was. I'm assuming the reader is at least vaguely familiar with the papers Lewis (1986a) and Lewis (1994).

I believe I can offer some at least semi-original thoughts on a few subjects related to the Principle; due to the size of the literature I cannot be sure that the concerns I raise have not been mentioned by someone somewhere, but I do have confidence that my take on them is novel. I will try to convince the reader of the following:

- that the problems regarding the notion of "admissibility" nullify the explanatory prospects of the project with regard to chance;
- that the "undermining problem" supposedly infecting Humean Supervenience is actually a modern take on an old, if not ancient, philosophical problem to which most philosophers already have their favorite solution;
- and finally, that someone who—following Lewis—thinks that the Principal Principle leads to contradiction in the presence of undermining futures can attempt to save the whole project by relaxing an assumption I call "chance-conditioning uniqueness", possibly concluding something about the Best System in the process: namely, that it is weaker than most debaters assume.

In this chapter I will attempt to be as formal as the authors involved in this subfield of philosophy, that is, almost not formal at all; like almost all the cited authors, I will simply assume that the structures and objects we are talking about exist, not because I believe it is the best way to go, but because I think I have something to add to the topic which is somewhat philosophical in nature and would not benefit from full formalization. It has to be noted that, finally, the topic of chance-credence norms has reached the level of mature formal discussion in the forms of papers like Bana (2016), Gyenis and Rédei (2016) and most importantly Gyenis and Rédei (2017).

Note that I am assuming that both the credence functions under discussion ("rational" or "reasonable" credences) and chance functions are probabilities, that is, that they conform to the axioms we discussed in the previous chapter. Lewis assumes reasonable credences are probabilities for "well-known reasons" (p. 98); I would like to point to those we have just seen as an example. That chances are probabilities supposedly follows from the Principal Principle; I will not discuss this here and will simply continue under this assumption. Whatever the interpretation of the chance function turns out to be, that is, whichever philosophical concept of chance we end up with while following the Lewisian project, it is—according to Lewis—supposed to conform to the classical probability axioms. Let us grant him this assumption.

I will also assume that propositions are sets of possible worlds. This is a view Lewis in general subscribed to (see, e.g., Lewis (1986b)); some might describe objects featuring in Lewisian approaches to attitudes *de se* as "relativized propositions" (see, e.g., Kallestrup (2012)), but to the best of my knowledge Lewis considered them objects of thought which are not propositions (sometimes, for example in Lewis (1979), calling them "properties"). It is true that in the "Subjectivist's Guide (...)" Lewis mentions the possibility of extending the approach so that an agent's "credence might be divided between different possibilities within a single world", but the only motivating example is that of "someone who is sure what sort of world he lives in, but not at all sure who and when and where in the world he is" (Lewis (1986a), p. 88–89); since we are not interested in problems of self-location in this book, I propose ignoring this. Let us then assume propositions are sets of possible worlds and continue.

3.1. Admissibility—first observations

Consider the following formulation of the Principal Principle (Lewis (1994), p. 483, formalism somewhat amended):

> Let Cr be a rational credence function for someone whose evidence is limited to the past and present. (...) Let ch be the function

that gives the present chances of all propositions.[2] Let A be any proposition. Let E be any proposition that satisfies two conditions:

- it specifies the present chance of A in accordance with ch;
- it contains no "inadmissible" information about future history; that is, it does not give any information about how chance events in the present and future will turn out. (We don't assume E is known.) (...)

Then the Principal Principle is the equation

$$\mathrm{Cr}(A|E) = \mathrm{ch}(A). \tag{3.1}$$

(We assume that some moment t is fixed, and the chance function "at t" is considered; we suppress the reference to time notation-wise, but we need to note its presence, since using terms like "the present" would otherwise make little sense.)

The Principal Principle will offer some message about objective chance if we understand what "inadmissible" really means. First, it seems natural to think of E as being a conjunction of two parts: one that specifies that $\mathrm{ch}(A) = x$ for some number x, and another, let's call it G, which may concern anything whatsoever, inasmuch as the information it carries is "admissible" in the above sense. (The tacit assumption is, of course, that the information that $\mathrm{ch}(A) = x$ is admissible; this is the root of the problem we will be discussing in the next section.) The Principle can then be restated, after repeating all the previous assumptions, as

$$\mathrm{Cr}(A|\mathrm{ch}(A) = x \wedge G) = x. \tag{3.2}$$

Second, applying the Principal Principle with an empty G leads, of course, to

$$\mathrm{Cr}(A|\mathrm{ch}(A) = x) = x. \tag{3.3}$$

So, it would seem that we arrive at some conclusion about inadmissible propositions: if G is one, then it might very well happen that for some rational credence function Cr and some proposition A,

$$\mathrm{Cr}(A|\mathrm{ch}(A) = x \wedge G) \neq \mathrm{Cr}(A|\mathrm{ch}(A) = x).$$

[2] Note that Lewis (1986a) left an option for the domain of chance functions not to include *all* propositions.

Note the similarity of these issues to the phenomenon of screening-off[3]:

Definition 10 (Screening off): *Assume a probability space* $\langle W, \mathcal{F}, P \rangle$ *is given. A proposition* C screens off *propositions* A *and* B *if and only if*

$$P(AB|C) = P(A|C) \cdot P(B|C) \tag{3.4}$$

or, equivalently,

$$P(A|C) = P(A|BC). \tag{3.5}$$

(The equivalence holds, of course, only if all the conditional probabilities are defined.)

Notice that screening off does not seem to favor any interpretation of probability. Subjectivists can read formula (3.5) as saying "given the information that C, gaining additional information that B does not change the credence in A". Frequentists of the finite persuasion will read it as saying that a relative frequency in a sequence is preserved when considering one of its subsequences. Hypothetical frequentists will say that what is preserved is the limiting frequency as sequence length goes to infinity. I am confident that propensity aficionados, once they figure out what to do with Humphrey's Paradox (for a relatively modern summary of the issue and some attempts at solutions see Humphreys (2004)), will also have a natural reading for the condition. (A promising newer attempt at a propensity interpretation of conditional probabilities is offered in Drouet (2011).) The point is that screening-off is a purely formal condition. Once we know that a function P defined on propositions is a probability measure, we can pick any three propositions and receive an answer to the question whether the first screens off the second from the third. No reference to any interpretation is needed.

We are, as already mentioned, operating under the assumption that rational credences are probabilities. Therefore, we can at least formulate a sufficient formal condition for inadmissibility: if for some rational

[3] Masterton (2010) has studied screening-off in the context of the Principal Principle, but in my opinion his research takes him in a different direction.

credence Cr and for some time t, G is not screened off from A by the proposition specifying the chance of A, that is, if

$$\mathrm{Cr}(A|\mathrm{ch}(A) = x \wedge G) \neq \mathrm{Cr}(A|\mathrm{ch}(A) = x),$$

then G is inadmissible w.r.t. A at time t. Note that Lewis writes "Admissible propositions are the sort of information whose impact on credence about outcomes comes entirely by way of credence about the chances of those outcomes" (p. 92). So, it seems we are on the right track: if the mentioned screening-off does not hold, then what happens is exactly that some rational credence in A is informed not only about the credence in the chance of A, but also about something else, namely, by G.

We could take these matters further and posit that the universal holding of the screening-off of the sort just mentioned was *definitional* of admissible propositions: that is, a proposition G is admissible w.r.t. A at time t if and only if any proposition to the effect that $\mathrm{ch}(A) = x$ for some real number x screens off G from A according to any rational credence Cr. (Obvious questions regarding the relative positioning of time and quantification appear; Lewis himself did not seem to be troubled by this.)

If the class of all propositions was given, and if the probability functions on them which were rational credence functions were specified, we would seem to have a clear—and, as a bonus, somewhat formal—notion of admissibility, which would be a welcome change from the literature which abounds in examples involving seers and crystal balls.

However, it would then seem doubtful whether the Principal Principle can do the job of "capturing all we know about chance" (p. 86), or even perform a seemingly lesser task of elucidating the concept of objective chance by invoking subjective probability (if you recall the title of Lewis's original paper on the topic). To understand the scope of the Principal Principle, to know for which propositions and credence functions the equation (3.2) holds,[4] we need to know which propositions are admissible; but to know which propositions are admissible,

[4] Yes, *a priori*, it may hold also in cases which in which the assumptions of the Principal Principle are violated; I propose noting this worry here but not elaborating on it.

we need to know all the true instances of (3.2). Therefore, appealing to the notion of screening-off will not help us here.

A different route would be to reformulate the Principle so that it would not involve any other propositions apart from A and the one specifying the chance of A, effectively culminating in equation (3.3); this would make the Principal Principle a variant of the so-called Miller's Principle (Miller (1966); note 1) that Miller tried to argue *against* it, not propose it, and 2) there are interesting issues regarding rigidity of designation and Miller's argument discussed in Jeffrey (1970)). However, this would certainly narrow the scope of the Principal Principle as intended by Lewis, who claims we have at least some intuitions about admissibility (p. 92–96: propositions about the past, propositions about how chance depends on history, and Boolean combinations of these are to be admissible).

There might be other ways to elucidate the notion of admissibility in a somewhat rigorous way (see, e.g., Meacham (2010) for a proposal which makes essential use of propositions expressing "theories of chance", which I will mention later). The spirit of the previous paragraph is not meant to suggest that there is some conceptual vicious circle in Lewis's view, but rather that finding the proper place for the notion of admissibility might lead to similar methodological issues as those regarding the concept of "natural properties", as extensively discussed in Lewis (1983): the notion appears in explications of notions useful in explicating it.

3.2. "Aboutness" and undermining

The issue of "undermining futures" is one that supposedly plagues accounts which attempt to analyze the concept of chance in a framework subscribing to Humean Supervenience: the claim "that the whole truth about a world like ours supervenes on the spatiotemporal distribution of local qualities". More explicitly, "all else supervenes on the spatiotemporal arrangement of local qualities throughout all of history, past and present and future" (Lewis (1994), p. 474). In my opinion whether we take it to be a substantial problem may depend on our view regarding what it means for a proposition, i.e., a set of worlds, to be *about something*.

Lewis gives the following characterization of a proposition being about a subject matter: "A proposition is about a subject matter—about history up to a certain time, for instance—if and only if that proposition holds at both or neither of any two worlds that match perfectly with respect to that subject matter" (Lewis (1986a), p. 93). Since we will never be in position to know exactly which worlds are possible, we should not expect to be able to apply this criterion with full confidence; however, it gives us some argumentative direction.

Consider the following consequence of this view of "aboutness". Suppose one subscribed to the following view, which might be seen as a form of determinism: if the arrangement of local qualities coincides at two possible worlds at some time, it also coincides at all other times. (This might be reasonable for example for those who take seriously talk of the future fundamental physical Theory of Everything being deterministic.) Suppose we grant the assumption of Humean Supervenience and restrict our attention to the set of possible worlds in which there are no two worlds with exactly the same arrangement of local qualities throughout all history (any difference between such worlds would be non-Humean). Then, colloquially, all propositions are propositions about the future. That is, for any world w and time t, the class of "worlds which match perfectly with w with respect to the arrangement of local qualities at all times later than t" is just $\{w\}$. So, if we want to subscribe to Humean Supervenience, and would like to claim that some propositions are about the past but not about the future, we cannot be determinists in the above sense: that is, we have to allow for the possibility that some worlds initially differ in the arrangement of local qualities but "converge" to identity in that respect.

(There might be various interesting notions of "aboutness" of propositions in the neighborhood of the one Lewis proposed, based on the fact that some regions of some worlds might match in some sense. However, it appears to me that any such notion is doomed to be problematic, since, for all we know about what's possible and not, it might very well be that in any possible world in which Caesar was killed by Brutus a specific configuration of sunspots materializes on the 15th of

May 2000, and so any treatise on the history of the Roman Empire will turn out to be also about some aspects of astronomy.

Interestingly, a rigorous sort of the "branching" approaches to possible worlds (approaches which Lewis vehemently opposed in his 1986 book)—the "Branching Space-times" framework by Nuel Belnap—features a similar problem, called in that context "funny business" (see, e.g., Belnap (2003).)

It should also be noted that (as Lewis eventually realized) propositions specifying the chance of some proposition at some time t may be not about the past, that is, about the (arrangement of local qualities) at times earlier than t, but about the future; specifically, for *some* propositions A, worlds w and times t, *some* propositions about chances at t of A are about the spatiotemporal arrangement of local qualities in world w at some times t' later than t. Suppose at t a coin is about to be tossed by a machine set to a specific angle and force. After the coin begins its flight through the air a magnet might be repeatedly turned on and off nearby in a very specific pattern which influences the result in a rigorous way (see Diaconis et al. (2007)). The chance at t of the coin landing heads up is determined, therefore, not only by the physical properties of the coin together with the "usual" physical features of the environment (gravity, air density, etc.), but also by whether the magnet is turned on or not during the coin's flight. There are therefore two chance propositions of interest: $CH1$, to the effect that $ch_t(Heads) = \alpha$, which is true in all worlds in which the magnet is not turned on, and $CH2$, to the effect that $ch_t(Heads) = \beta$ for some $\beta \neq \alpha$, which is true in all worlds in which the magnet is turned on. And so $CH1$ and $CH2$ are chance propositions, specifying the chance of some proposition at time t, which are nonetheless about the future, that is, about times later than t.

With this in mind let us look at how Lewis introduces the problem of "undermining futures" in (Lewis (1994), p. 482–483):

> [S]uppose we have a Humean analysis which says that present chances supervene upon the whole of history, future as well as present and past. (...) Then different alternative future histories would determine different present chances. (...) And let's suppose, further, that the differences between these alternative futures

> are differences in the outcomes of present or future chances events. Then each of these futures will have some non-zero present chance of coming about.
>
> Let F be some particular one of these alternative futures: one that determines different present chances than the actual future does. F will not come about since it differs from the actual future. But there is some present chance of F (...) some present chance that events would go in such a way as to complete a chancemaking pattern that would make the present chances different from what they actually are. The present chances *undermine* themselves. (...)
>
> [considering some future different from the actual one] Could it come to pass, given the present chances? Well, yes and no. It could, in the sense that there's non-zero present chance of it. It couldn't, in the sense that its coming to pass contradicts the truth about present chances. If it came to pass, the truth about present chances would be different. Although there is a certain chance that this future will come about, there is no chance that it will come about while still having the same present chance it actually has. It's not that if this future came about, the truth about the present would change retrospectively. Rather, it would never have been what it actually is, and would always have been something different.
>
> This undermining is certainly very peculiar. But I think that, so far, it is no worse than peculiar.

Note the last bit: Lewis does not consider undermining to be a big problem in itself. The real issue, the Big Bad Bug, is a supposed contradiction resulting from the application of the Principal Principle, to which we will turn in the next section. All the same, I would like to claim that the undermining phenomenon is actually a well-known—ancient, actually—philosophical problem in disguise, and is not peculiar at all in its own way.

Consider the "fair coin toss + magnet" scenario we have just discussed. Consider chances at t, before the toss, which takes place at some moment $t' > t$. Introduce an additional random element: the magnet is switched on if and only if a certain radium atom decays in the interval between t and t'. In the actual world the magnet is not turned on and, since we are trying to be Humean, we should say that the chance at t of the coin ending "heads up" is 1/2: chances supervene on the past, present and future, and not on what goes on in other worlds. However, *it is possible* that the radium atom decays between t and t′. If it indeed decayed, the magnet would be turned on, and the chance at t of the coin ending "heads up" would be something other than 1/2. But notice that we have changed the possible world we are talking about: it should not be surprising that if the chancemaking pattern after t differs between the actual world and some possible world, chances at t also differ.

Let us add some more detail to the example: in the actual world the coin falls "heads up" and, if the magnet is turned on, it falls "tails up" (see again Diaconis et al. (2007) for a discussion of such setups). Suppose t is the present time. In the actual world it is presently true that the coin will fall "heads up". Is it possible for the magnet-future to realize? Well, it is, since the radium atom *can* decay between t and t′; there's a non-zero present chance of it. However, *if it* decayed, the truth about the present truths would be different: for example, it would be false that the coin will fall "heads up". It is not possible for the magnet-future to realize without the change of truth values of statements about the future at t.

What I wish to say is that the "undermining futures" issue is simply another rendition of the ancient problem of future contingents. Since we have already seen that, for Humeans at least, some chance propositions are about the future, and since there are non-zero chances of various futures being realized, so it is possible for various futures to be realized and Humeans should accept that *chance statements may be future contingents.* There is a plethora of views regarding these: note, for example, that for Lewis himself all future contingents are true or false; for an overview of various positions see Øhrstrøm and Hasle (2015). It seems to me that a wise move would be to choose one of the general approaches to the issue, for example the assessment sensitivity idea from MacFarlane (2014), and then see whether it ties properly with how one would like to conduct one's Humean research: there is nothing particularly peculiar to the "undermining futures" issue. One does not need even to follow in Lewisian footsteps regarding one's view of possible worlds; instead, a branching view could be combined with a Humean approach to chances and a version of the Thin Red

Line position with regard to future contingents (see, e.g., Malpass & Wawer (2012)).

However, there is a *different* problem to be tackled if one tries to combine Humean Supervenience with the Principal Principle: it is the "Big Bad Bug", as it is popularly called, and is the one to which we now turn.

3.3. The Big Bad Bug: the Best System may be weaker than you think

To get a grasp on the issue look, first, at how Lewis continues the quote we used on p. 65 when introducing the Principal Principle:

> (...) the Principal Principle is the equation
>
> $$\mathrm{Cr}(A|E) = \mathrm{ch}(A). \tag{3.1}$$
>
> Now take A to be F, our alternative future history that would yield present chances different from the actual ones; and let E be the whole truth about the present chances as they actually are. We recall that F had some present chance of coming about, so by the Principal Principle, $\mathrm{Cr}(F|E) \neq 0$. But F is inconsistent with E, so $\mathrm{Cr}(F|E) = 0$. Contradiction. I could tolerate undermin[ing] as merely peculiar. But not contradiction! (Lewis (1994), p. 483)

The Big Bad Bug, as Lewis himself called the problem, is this contradiction: the Principal Principle seems to be incompatible with Humean Supervenience.

The reader will, of course, be perfectly excused if (s)he thinks that a serious discussion of anything resembling a formal contradiction has to proceed beyond the vague talk displayed in and around the quotes in this chapter. In that case I suggest exploring the already cited papers by Rédei, Z. Gyenis and Bana.

It has to be mentioned that some philosophers have thought that there is no need for any "debugging"; see, e.g., Roberts (2001). It seems, though, that most, including Lewis himself, believed that some modification of the Principal Principle was in order (from the vast literature on the subject I suggest starting with Hall (1994)). While I will also suggest a modification of the Principle, it will be motivated by a suggestion regarding the so-called "Best System" approach to lawhood—another famous idea by Lewis which can be tied to a reformulation of the Principal Principle he proposed.

According to the Best System approach, to quote Cohen and Callender (2009) (p. 2), "the laws of nature are the true generalizations that best systematize our scientific knowledge"; a bit more explicitly, "a true generalization is a law if and only if it is an axiom of all the 'Best Systems'—axiomatic systematizations that best balance strength and simplicity" (p. 5). Lewis himself in his 1994 paper uses the singular, saying that "[t]he best system is the one that strikes as good a balance as truth will allow between simplicity and strength. How good a balance that is will depend on how kind nature is. A regularity is a law iff it is a theorem of the best system" (p. 478). We will ignore the issue of how many Best Systems there are in what follows. The important thing is that Lewis stresses that the best-system analysis is "Humean" (p. 480). I will suggest a way in which one can coherently subscribe to Humean Supervenience, the best-system analysis, and a version of the Principal Principle, on pain of believing that the Best System does not provide theorems as strong as, I think, is commonly presumed.

Are propositions about chances theorems of the Best System? Not according to the initial conception, since there the propositions making up the system are required to be true, and before an analysis of chance is proposed, it would be premature to set the same requirement on propositions of this type. Lewis adds, then, another balancing factor to simplicity and strength: the one of "fit". The higher the fit of a system, the higher the chance of the actual future according to the system (Lewis (1994), p. 480). The Best System is the one which displays the best balance of strength, simplicity, and fit. Lewis can then say that "the chances are what the probabilistic laws of the best system say they are" (*ibid.*).

Lewis explicitly assumes that some systems say "what the chances will be when situations of a certain kind arise" (*ibid.*). In the earlier "Subjectivist's Guide (...)" he considers a version of the Principal Principle in which after the conditioning sign a specific kind of admissible propositions is featured, namely "history to chance conditionals".

Among others, they should satisfy these two requirements: "(1) The consequent is a proposition about chance at a certain time. (2) The antecedent is a proposition about history up to that time; and further, it is a complete proposition about history up to that time" (Lewis (1986a), p. 95); they *might* be systematic, "compressible into generalizations" to be thought of as probabilistic laws (p. 96). With all this in mind we can now cite one of Lewis's reformulations of the Principal Principle:

> Given a time t and world w, let us write ch_{tw} for the chance distribution that obtains at t and w. For any proposition A, $\mathrm{ch}_{tw}(A)$ is the chance, at time t and world w, of A's holding. (The domain of ch_{tw} comprises those propositions for which this chance is defined.) Let us also write H_{tw} for the complete history of world w up to time t: the conjunction of all propositions that hold at w about matters of particular fact no later than t. H_{tw} is the proposition that holds at exactly those worlds that perfectly match w, in matters of particular fact, up to time t.
>
> Let us also write T_w for the complete theory of chance for world w: the conjunction of all the conditionals from history to chance, of the sort just considered, that hold at w. Thus T_w is a full specification, for world w, of the way chances at any time depend on history up to that time.
>
> Taking the conjunction $H_{tw}T_w$, we have a proposition that tells us a great deal about the world w. It is nevertheless admissible at time t, being simply a giant conjunction of historical propositions that are admissible at t and conditionals from history to chance that are admissible at any time. (...) Therefore we have:
>
>> **The Principal Principle Reformulated.** Let Cr be any reasonable initial credence function. Then for any time t, world w, and proposition A in the domain of ch_{tw}
>>
>> $$\mathrm{ch}_{tw}(A) = \mathrm{Cr}(A|H_{tw}T_w). \quad (3.6)$$
>
> In words: the chance distribution at a time and a world comes from any reasonable initial credence function by conditionalizing on the complete history of the world up to the time, together with the complete theory of chance for the world. (p. 96–98, some changes in notation)

On the one hand, this looks to be a weaker principle than the one quoted earlier: it restricts our attention to a specific type of proposition to the right of the conditioning sign.[5] On the other hand, I suggest that it is still very strong, and some weakenings of it can be suggested which avoid the Big Bad Bug and are still able to say something meaningful about the relationship of chance and rational credence. However, I will say upfront that significant doubts about the rationality of the principles I will propose can certainly be formulated.

The main observation is the following: considering the rational credence in some proposition A, both the original Principle and the just cited rephrasing formulate a conditional constraint given a proposition which specifies the chance of A uniquely. Let me call this feature of the principles *chance-conditioning uniqueness*. If this is relaxed, the contradiction may be avoided. There are many intuitive weak principles of this kind; for example "given the information that the chance of A is in the $[0.4, 0.7]$ interval, a credence of 0.1 in A would be irrational". In the remaining part of this chapter I will sketch a few proposals for how a version of the Principal Principle formulated in this spirit might look. I am not worried too much that these proposals have clearly visible flaws since I believe a serious discussion of these topics should eventually move to the framework of higher order probability spaces (see the already mentioned Gyenis and Rédei (2017) and the next chapter).

Let us set aside the issue of admissibility of the history-to-chance conditionals and concentrate on what the conditionals say. Notice that the only requirement Lewis proposes for their consequent is, as already mentioned, that it "is a proposition about chance at a certain time". The contradiction in the Big Bad Bug is reached, remember, because a non-zero chance is appropriated to an inconsistent proposition. However, there is no need for a true history-to-chance conditional to specify the chance exactly; not only could it be given as belonging to an interval ("if the experimental setup is prepared so-and-so, the chance of obtaining this particular measurement result is 0.3 ± 0.00005"),[6] but

---

[5] Lewis himself treated (in the "Subjectivist's Guide (...)") the two formulations as equivalent, modulo a few quibbles.

[6] I am not suggesting that chances are unsharp, but there are best-system-style analyses which do, see, e.g., Dardashti et al. (2014).

also to some discrete set ("since, in the future, before the measurement an intervening factor may come into play, the current chance that in the future this particular measurement result will be obtained is either 0.3 or 0.7").

Consider, then, the above quote regarding "Principal Principle Reformulated". Change the beginning of the second paragraph so that it does not describe T_w, and instead speaks about T_{tw}:

> Let us also write T_{tw} for the complete theory of chance for world w up to time t: the conjunction of all the conditionals from history up to time t to chance that hold at w. Thus T_{tw} is a full specification for world w of the way chances at time t depend on history up to that time.

It might very well happen, as it does for example in Lewis's tritium example, that for some A the proposition $H_{tw}T_{tw}$ does not entail that the chance of A equals some single specified value. Since the main intuition behind the discussed principles is that rational credence should conform to credence about chance, we should expect that on the basis of that proposition rational credence is constrained, but sometimes not uniquely determined, which the following principle tries to capture[7]:

> **A Debugged Reformulated Principal Principle (DRPP).** Let Cr be any reasonable initial credence function. Then for any time t, world w, and proposition A in the domain of ch_{tw},
>
> $$\mathrm{Cr}(A|H_{tw}T_{tw}) \in \{ch_{tv}(A) \mid v \text{ is indistinguishable from } w \text{ up to time } t\}. \tag{3.7}$$

> In words: the chance distribution at a time and a world constrains what is obtainable by conditionalizing a reasonable initial credence function on the complete history of the world up to the time, together with the complete theory of chance for the world up to that time.

---

7 This perhaps comes close to the "loophole" Ismael (2008) mentions in her footnote 9 but decides not to explore.

While it is clear to me that if we stay on the vague level of discourse particular to the mainstream discussions about the Principal Principle, the DRPP avoids the Big Bad Bug, it certainly gives rise to peculiar issues. As will be the case with other variants discussed below, they are most easily seen when reading the Principle "from left to right"; that is, for example, asking the question "what is the rational credence in A given $H_{tw}T_w$?" The "$\in$" sign in the DRPP can be interpreted in at least two ways. One of them would say that there are as many $Cr(A|H_{tw}T_w)$'s as there are $ch_{tv}(A)$'s—that any chance function compatible with the given information about the past gives rise to a rational credence. However, given that the conditionalization operation is a function, that is, it uniquely determines the posterior given the prior, it would follow that there are nontrivially different unconditional rational credence functions. While this is by no means absurd, it is perhaps unfortunate and unexpected that it is a consequence of what is to be a principle connecting credence about chance to credence.

Another reading of the "$\in$", somewhat weaker, would have the DRPP state that *the* rational credence in A given $H_{tw}T_{tw}$, whatever it is, belongs to the set of ch_{tv}'s for v indistinguishable from w up to time t. This avoids the previous problem, if it is a problem at all, but some may find it to be too weak if the basis for excluding some ch_{tv}'s as irrational values for the credence in A is not given (and I will hazard no proposals for it).

Perhaps a fruitful direction to pursue would be to consider chance-credence principles not as concerning rational credence, but as credence it is epistemically permissible to have (for an interesting discussion of the two notions see, e.g., Hughes (2017), although the author defends the claim that in the case of full belief, the notions are coextensive). The principles violating chance-conditioning uniqueness could be seen, then, as describing how chance constrains the set of epistemically permissible credences.

This, however, does not solve a different issue. Suppose there are two values of the chance of A compatible with the given past; take the "coin + magnet" example with two different chances, α and β, depending on whether the magnet is turned on in the future. Why should we insist that before the toss and before the state of the magnet is decided the rational credence in A given all information about the past and the complete theory of chance of the given world up to that time should belong to the set $\{\alpha, \beta\}$? Choosing one of these over the other would seem arbitrary, so we might be tempted to say that it is epistemically permissible to hold any of the two (and, to reiterate, the value of the DRPP for the analysis of chance lies in its description of how chance constrains the sets of epistemically acceptable credences). But this disregards the prior credences about the chances of each future actualizing. It might very well be that the agent has learned, e.g., something about the way the magnet is operated which leads her to the credence 0.9 that it will be turned on, that is, that the future will be such that the chance of A is β. If the fact that the agent obtains a rational credence about the chance of the magnet being turned on in the future is already inside H_{tw}, this is especially troubling. This suggests that the set of epistemically permissible credences in A given $H_{tw}T_{tw}$ should contain at least some convex combinations of the possible chance values.

This—that is, not only removing the requirement of chance-conditioning uniqueness, but also taking into account the prior credences in which of the possible futures will actualize—seems to take us (more in spirit than in the letter) in the direction of the General Recipe of Ismael (2008) (p. 298). Instead of trying to work out the details of the connection, which I'd consider fruitful in a more formal framework of higher-order probability spaces, I would like to propose a different approach to what is conditionalized upon in the variations of the Principal Principle one can propose to avoid the Big Bad Bug: we could turn to what is offered by the Best System.

The "complete theory of chance" (whether "up to some time" or without that qualification) is most likely not a part of the best system—it deals with, recall, "complete propositions about history" up to a certain time. If this was a part of the Best System, then at least when it comes to chance the system would seem to peculiarly favor strength over simplicity. Lewis himself, in the postscript to the "Subjectivist's Guide (...)", made the decision not to require that chances should be completely governed by laws, requiring only that the "history-to-chance conditionals will not conflict with the system of laws of chance" (p. 126–127). However, we do not need to impose such a strict requirement; rather we only need to formulate a restricted version of the Principal Principle where what is given to the right of the conditioning bar is the information from the Best System together with the historical data which make the former applicable.

The resulting principle, while certainly weaker than the original, would have an advantage in that we know that however we think of the language of the eventual Best System—be it fundamental physics or something else (see Frisch (2014))—we can be reasonably sure that people are working on (something leading towards) it. This makes it at least *prima facie* reasonable to formulate a constraint on rational credence that ties the credence about chance to what the Best System says about chance. In contrast, I think it is fairly certain that no one is pursuing the goal of providing a complete history of any possible world or a complete theory of chance (in the Lewisian sense) for it (even up to a certain time). However, it is not evident that the issues of practicality (for lack of a better term) should be in the spotlight in the context of metaphysics.

Let $\mathrm{BS}(t, w, A)$ be a set of numbers: the possible values $\mathrm{ch}_{tv}(A)$ might have according to the Best System given that the past of v and w are indistinguishable from the point of view of the Best System; that is, there is no theorem of the Best System specifying a difference in the arrangement in local qualities in w and v before t. For example, in our "coin + magnet" case, if w is the actual world and v the one in which the magnet is turned on in the future, it might very well be that $\mathrm{BS}(t, w, A) = \{\alpha, \beta\}$. Note the "might": we need to be careful when talking about the Best System since we do not have a clear idea of what it will be. I am fine with this since I want to sketch a position which a Humean can coherently maintain while avoiding the Big Bad Bug, and for this the particulars of the Best System are of little importance. However, I need to note that for example Loewer (2004) decided to boldly claim that "there are coherent credence functions that violate the Principal Principle for L-chance", where L-chances are chances as given by the laws of the Best System; this without knowing what the laws are, and so, it would seem, without being able to produce any example of such a violation. The set $BS(t, w, A)$ might not exist at all, if for example the proposition A is chosen so that the Best System is silent about its chances. (One could define $BS(t, w, A)$ to equal $\emptyset$ in such a case, of course.)

Let me now present the idea behind another version of the Principal Principle: the **Best System Principal Principle** (BSPP). Chance constrains credence (at least) as follows. A rational credence in A, given what the Best System says about the chance of A, is consistent with that information: if the Best System specifies a single value, then the rational credence is that value, while if the Best System specifies a set of values, then these are precisely those which are epistemically permissible at that time (given what the Best System says, remember: additional information may change this). That is:

> **The Best System Principal Principle (BSPP).** Let Cr be any reasonable initial credence function. Let BS be the part of the Best System containing laws governing chance and let H_{tw} be the information about world w and its history up to time t from which, together with BS, there is a derivation of what the set $BS(t, w, A)$ is. Then
>
> $$Cr(A|H_{tw}BS) \in BS(t, w, A). \qquad (3.8)$$

A few additional issues crop up, of course. One of them is whether the applicable part of the Best System is equivalent to a single proposition. A more serious one concerns the actual strength of this principle, namely, how big a constraint on rational credence it expresses and how successful it is in avoiding the Big Bad Bug. It might be that some examples of undermining in which an application of the original Principle leads to the Bug involve situations which fall under the laws forming the Best System; the more the better for the BSPP. But, since we do not know what the Best System is, we cannot exclude the possibility that most—if not all—derivations of the contradiction which forms the Big Bad Bug involve matters on which the Best System is silent, which would make the BSPP quite toothless.

Coming back to one previously mentioned problem, note that in our "coin + magnet" example if the Best System entails that, say, $\alpha = 0.3$ and $\beta = 0.7$, it does not follow that the rational credence is 0.5, or any number different from 0.3 or 0.7, or, in fact, that it is any particular number at all. The Best System may say "the laws dictate that there are two chances of A compatible with the past, 0.3 and 0.7", and this simply underdetermines rational credence, or even epistemically permissible credence. So, just like before, where instead of the Best System we discussed complete theories of chance for a world up to some time, perhaps another version of the principle is required, this time accepting as epistemically permissible values of the conditional credence in A some values of convex combinations of the elements of $BS(t, w, A)$.

While I meant it to be clearly visible that I do not have a version of the Principal Principle with which I would be fully happy, I hope I have convinced the reader at least of the potential for relaxing the chance-conditioning uniqueness assumption as one way of dealing with the Big Bad Bug.

Chapter 4

A few remarks on higher-order probabilities

I mentioned a few times in the previous chapter that in my opinion a serious discussion of the Principal Principle would best take part in some framework featuring higher-order probabilities. This idea—that is, the notion that it makes sense to say (and model formally) things like "the probability that the probability of an event A equals x equals y", or "the credence that the chance of A equals x equals y", if both credences and chances are probabilities—has been taken seriously in logic and computer science in recent decades. (For a classical account, see Fagin & Halpern (1994); for some recent takes on the topic, see for example De Bona et al. (2015) and Atkinson and Peijnenburg (2013) (for a description of infinite-order probabilities).) It seems that in formal epistemology the approach by Gaifman has gathered the biggest popularity. In his classic 1988 paper he defines a HOP to be a quadruple $\langle W, \mathcal{F}, P, Pr \rangle$, where $\langle W, \mathcal{F}, P \rangle$ is a classical probability space, Pr is a mapping associating with each pair $\langle$an element of $\mathcal{F}$, a Borel subset of $\mathbb{R}\rangle$ an element of $\mathcal{F}$, and the quadruple satisfies six axioms. The intended reading of, say, the expression $Pr(A, (0.5, 1])$ is "(the event that) the probability of A is higher than 0.5". Axiom VI, which I will not discuss here in detail, connects Pr and P. Not only does it bear some resemblance (noted by Gaifman himself) to the Principal Principle, but also its violation leads to Dutch-bookability. (One of the examples of this in the original 1986 version of the paper (p. 285) seems to me to be incorrect, while the 1998 version offers a long, carefully written out, and—to my eyes—correct argument: this is another reason for reading the 1998 version and not the reprint of the previous one in Arló-Costa et al. (2016).)

As I have already mentioned, Gyenis and Rédei (2017) (among others) study in detail the feasibility of various kinds of the Principal Principle in a formal framework. Speaking a bit informally, they investigate whether it is possible that for any Boolean algebra of random events there exists a Boolean algebra—on which the credence functions are to be defined—which would contain the previous one and also all propositions specifying the probabilities of those random events so that appropriate algebraic relationships hold which would ensure that the conditional probabilities in the given variant of the Principle are defined. This is at the core of one of the notions of consistency the authors are considering. However, at a certain point they discuss Gaifman's HOP framework and, indeed, formulate a concise open problem regarding the consistency of the Principal Principle: "whether any objective probability theory can be made part of a HOP [Higher-Order Probability Space] in such a way that the objective probabilities are related to the subjective ones in the manner of [Gaifman's Axiom VI]" (p. 21). They proceed to present the problem formally, but doing so here would require introducing the details of Gaifman's framework, which I will not do since I do not have a solution to that problem. I just want to direct the reader's attention to the fact that there is an open problem of immediate relevance to the possibility of formal argumentation about variants of Principal Principle in the esteemed framework of higher-order probability spaces: see Gyenis and Rédei (2017, p. 21).

In Section 4.2 I will investigate a greatly simplified but constructive version of the problem, now turning to an issue regarding the capturing of learning the chance of a proposition as Bayesian Conditionalization.

4.1. Updating on learning a chance of a proposition: a negative result

In a sense, Bayesian Conditionalization is a *general* method of updating, when you allow for expanding of the probability spaces involved. We will use the formal notion of "extension of a probability space" in the next section, but here it will suffice to say that one of the ways of thinking about it is that of "fine-graining": the extension, or expansion, of the original probability space "contains" the old one, but includes more propositions which result from dividing the old ones. For a pioneering result regarding the generality of Bayesian Conditionalization see Diaconis and Zabell (1982), and for a more generalized approach unifying many update methods consult Gyenis (2014)[1].

Consider now what happens when you learn (only) "that the chance of A is x". One might be tempted to think in the following manner:

- you should set your credence in A to x (possibly by adhering to some form of the Principal Principle);
- but doing *just that* will make you violate Probabilism, so set your credence in $\neg A$ to $1-x$ and update using Jeffrey Conditionalization[2];
- taking a cue from the Diaconis and Zabell result, we could believe it should be possible to capture all this using a single probability space containing both the proposition "A" and "the chance of A is x", so that in that space the update is just Bayesian Conditionalization (on "the chance of A is x").

But this turns out to be impossible due to the following result:

Fact 7 (B. Gyenis, personal communication[3]). *There exists no probability space $\langle W, \mathcal{F}, P\rangle$ for which the following principle holds:*

$$\forall H \in \mathcal{F} \qquad P(H|\text{“}Ch(A) = x\text{”}) = x \cdot P(H|A) + (1-x) \cdot P(H|\neg A), \qquad (4.1)$$

[1] We are preparing a substantially improved version, intending to publish it as a joint paper.

[2] If the Reader does not know this method, I advise consulting Section 5.1.2 below.

[3] B. Gyenis proved the result and informed me of it when we were working on Gyenis & Wroński (2017).

where A *and* "$\mathrm{Ch}(A) = x$" *are propositions in* $\mathcal{F}$ *satisfying* $0 < x, P(A) < 1$, $x \neq P(A)$.

(For an almost immediate proof of this fact, take $H = \neg$"$\mathrm{Ch}(A) = x$".)

Note that it is an elementary fact that, for any H, $P(H) = P(A) \cdot P(H|A) + P(\neg A) \cdot P(H|\neg A)$, and so, if the chance of A is x, and P is the credence of an agent after she learns this (which also means that the chance of $\neg A$ is $1 - x$ and the agent has learned this), we should expect P to satisfy—remember, for any H—the right-hand side of equation (4.1). However, due to Fact 7, in such a situation P *cannot* be the result of conditionalization on the proposition that "Ch(A)=x".

This might suggest that one cannot just take a proposition, slap a label on it with the inscription "I'm saying that the chance of A equals x" and expect it to do the work of a chance proposition when its truth is learned. More troublingly, due to the universal quantifier used the result suggests that *no* proposition is fit to fill this role!

One route which might suggest itself would be to use for example Gaifman's notion of a higher-order probability space, where chance propositions are not elements of the algebra on which the measure is defined with some more or less arbitrary labels, but rather the values of the function Pr, which has a rigorous connection with the probability measure P; the issue of capturing learning a chance proposition as conditionalization would take a different form in such a case and for example the Diaconis and Zabell result would not be expected to hold.

However, it is not evident that the message of Fact 7 is indeed so troubling as I just suggested.[4] To prove the contradiction, one has to use as H a proposition defined in terms of the proposition "Ch(A)=x" (at least, this is what I believe until someone finds a different proof). We may think of the belief update situation as combining two steps: first, an agent with the space of propositions $\mathcal{E}$ fine-grains it into $\mathcal{F}$ (again, for a formal discussion of this see the next Section), which now includes propositions about chance, and *then* conditionalizes on one of the "new" propositions. Equation (4.1) would then be expected to hold just for those elements of $\mathcal{F}$ which were already in $\mathcal{E}$. Perhaps, for example, a version of the above principle which constrains the application of the

[4] I would like to thank Ronnie Hermens for asking me about this.

equation (4.1) only to propositions which are Boolean combinations of propositions which are not about chance could be given a precise sense and be shown to be satisfiable. This is work for the future.

4.2. Higher-order probabilities as expert functions: a construction

I will now turn to a constructive contribution to the problem of extending a probability space with propositions which are *about* probability, in a minimal, precise sense I will soon introduce. Consider a finite probability space $\langle W, \mathcal{F}, P\rangle$ corresponding to credences of some agent $\mathcal{A}$. Suppose an agent $\mathcal{B}$ has her own view on the same matters (i.e., on $\mathcal{F}$), given by a probability function Q. Metaphysics enthusiasts or just those interested in the Principal Principle may think of a possible objective chance function instead of a credence function of a different agent. Similarly, belief update aficionados may think of the other agent as a future version of the first agent. (These options have already been suggested, for example in Gaifman (1986).) If $\mathcal{A}$ thinks that $\mathcal{B}$ is an expert when it comes to the propositions in $\mathcal{F}$, her conditional credence in a proposition A, given that $\mathcal{B}$ is right, should equal $Q(A)$. Is there an event in $\mathcal{F}$ which would play the role of "that $\mathcal{B}$ is right" (... "that the objective chance is given by Q", etc.), that is, an event E such that for any $A \in \mathcal{F}$, $P(A|E) = Q(A)$? Sometimes there is one—precisely in those cases in which Q is P conditional on some event in $\mathcal{F}$. But in general, such an event does not exist. We need to extend the original space so that apart from the propositions in $\mathcal{F}$ it contains additional material we can use to generate the relevant events.

Additionally, if we consider more experts, i.e., different functions Q_i, their verdicts will be incompatible, and so the events, say, "that $\mathcal{B}_1$ is right" and "that $\mathcal{B}_2$ is right", need to be disjoint. These two conditions I take to be minimal if we want to say that the resulting extended space corresponds to degrees of belief an agent has towards the propositions in $\mathcal{F}$ as well as towards propositions expressing the degrees of belief of some experts. We will be using the following notion:

Definition 11 (extension of a probability space): A probability space $\langle W', \mathcal{F}', P' \rangle$ is an *extension* of the probability space $\langle W, \mathcal{F}, P \rangle$ by means of a homomorphism ϕ if ϕ is a Boolean algebra embedding of $\mathcal{F}$ into $\mathcal{F}'$ such that for any $A \in \mathcal{F}$ $P'(\phi(A)) = P(A)$.

In other words, an extension of a given space preserves the "old" probabilities, while possibly introducing some new events which may correspond to new factors the agent has taken into account—in a sense the space becomes more fine-grained. In philosophy this idea has been frequently used, for example in the context of Reichenbach's Common Cause Principle (see the discussion in Wroński (2014)). Similar structures are used in the theory of stochastic processes and are known as "filtrations".

Using the above notion of extension we can now formulate and prove the following:

Theorem 4. *Given a finite probability space $\langle W, \mathcal{F}, P \rangle$ and probability measures $Q_1, \ldots, Q_n$ on $\mathcal{F}$ there exists a finite probability space $\langle W', \mathcal{F}', P' \rangle$ which is an extension of $\langle W, \mathcal{F}, P \rangle$ by means of a homomorphism ϕ and is such that for any $1 \leqslant i, j \leqslant n$ there exist $E_i, E_j \in \mathcal{F}'$ such that:*

- $\forall A \in \mathcal{F}\ P'(\phi(A)|E_i) = Q_i(A)$;
- *if* $i \neq j$, *then* $E_i \cap E_j = \emptyset$.

Proof. The construction of the extension is given as Algorithm 1 (p. 92). The algorithm requires $\mathcal{F} = 2^W$ but is easily adaptable to other cases.

Line 2 ensures that the image of ϕ is a subalgebra of $2^{W'}$ isomorphic to 2^W. Due to lines 7 and 11 it holds for every atom of the original space, that is, a singleton $\{w^i\}$ for some $1 \leqslant i \leqslant |W|$, that $P'(\phi(\{w^i\})) = P'(\bigcup_{0 \leqslant k \leqslant n} \{w^i_k\}) = \sum_{0 \leqslant k \leqslant n} P'(\{w^i_k\}) = P(\{w^i\})$, therefore the new space is an extension of the old one. Line 4 guarantees that for $i \neq j$ $E_i \cap E_j = \emptyset$. That each event E_i plays the role summarized by "probability on the original algebra is given by Q_i", that is, that for any $A \subseteq W$ $P'(\phi(A)|E_i) = Q_i(A)$, is guaranteed by line 7.

All the above is made possible by line 5, which is the key step in the construction. There we put $P'(E_k)$ as the highest fraction $\frac{1}{2^m}$ (for $m \geqslant n$) which can be distributed among the singletons of elements of

E_k in such a way that (*) E_k plays the expert role w.r.t. the atomic events of the original algebra 2^W (the left conjunct) and (**) the probabilities of these singletons will be low enough that satisfying the extension requirement remains possible (the right conjunct). That such a number exists is a matter of elementary algebra—the probability of E_k can be put arbitrarily low. Each atom of the original space, $\{w^i\}$, corresponds (via ϕ) to the event $\{w_0^i, \ldots, w_n^i\}$. The right conjunct in line 5 ensures that we set the probabilities of $P'(\{w_1^i\})$, $P'(\{w_2^i\})$ and so on such that their sum will not exceed $P(\{w^i\})$—and the remainder is set in line 11 to be $P'(\{w_0^i\})$.

The algorithm defines P' on all atoms of $2^{W'}$, which, since in this chapter we are dealing only with classical probability, is equivalent to defining the whole P' as a measure on $2^{W'}$. □

Notice that if Q_i is an expert function for P and that in the extended space E_i is the event "the expert *i* is right"; that is, if $P'(\phi(A)|E_i) = Q_i(A)$, then

$$P'(E_i) = \frac{P'(\phi(A) \wedge E_i)}{Q_i(A)} \leqslant \frac{P(A)}{Q_i(A)}.$$

That is, the P and each Q_k impose an upper bound on the probability of E_k. This bound makes some sense when P is interpreted as some agent's subjective probability: if I believe some A to a very low degree, then presumably I will not have a high degree of belief that an expert who believes it to a very high degree is correct. This has to be taken with a grain of salt—the upper bound on $P'(E_i)$ is the minimal of *all* fractions $\frac{P(A)}{Q_i(A)}$ for any A. Therefore, we see that some degree of coherence is built into the extended structure, although no "symmetrical" lower bound is imposed.

4.2.1. Questions

I have shown that Algorithm 1 produces extensions of probability spaces that satisfy two conditions which are minimal if one wants to say that these extensions result from adding expert functions (or objective chance functions, etc.) to the initial space. It might be interesting to consider some refinements or possible outright improvements of the algorithm.

Algorithm 1. Extend a given probability space as specified in Theorem 4.

Require: A nonempty set $W = \{w^1, \ldots w^{|W|}\}$ and probability measures $P, Q_1, \ldots, Q_n$ on 2^W.

Ensure: A probability space $\langle W', 2^{W'}, P' \rangle$ which is an extension of $\langle W, 2^W, P \rangle$ by means of a homomorphism ϕ and is such that for any $i, j \in \{1, \ldots, n\}$ there exist $E_i, E_j \in 2^{W'}$ such that:

- $\forall A \in 2^W \; P'(\phi(A)|E_i) = Q_i(A)$;
- $E_i \cap E_j = \emptyset$.

$W' \leftarrow \{w^i_k \mid 1 \leqslant i \leqslant |W|, 0 \leqslant k \leqslant n\}$

$\phi \leftarrow \{\langle A, \cup_{0 \leqslant k \leqslant n, w^i \in A} \{w^i_k\} \rangle \mid A \subseteq 2^W\}$

for $k = 1$ **to** $k = n$ **do**

    $E_k \leftarrow \cup_{0 \leqslant i \leqslant n} \{w^i_k\}$

    $P'(E_k) \leftarrow \max \left\{ \frac{1}{2^m} \mid m \geqslant n, \exists \{c^i\}_{1 \leqslant i \leqslant |W|} \forall_{1 \leqslant i \leqslant |W|} \left(c^i = Q_k(\{w^i\}) \cdot \frac{1}{2^m} \wedge c^i < \left(P(\{w^i\}) - \sum_{1 \leqslant l < k} P'(\{w^i_l\}) \right) \right) \right\}$

    for $i = 1$ **to** $i = |W|$ **do**

        $P'(\{w^i_k\}) \leftarrow Q_k(\{w^i\}) \cdot P'(E_k)$

    end for

end for

for $i = 1$ **to** $i = |W|$ **do**

    $P'(\{w^i_0\}) \leftarrow \left(P(\{w^i\}) - \sum_{1 \leqslant k \leqslant n} P'(\{w^i_k\}) \right)$

end for

First, one might want to impose some other bounds on $P'(E_k)$ that derive from the features of P and Q_k apart from the upper bounds described at the end of the previous subsection. Perhaps if the minimal absolute value of the differences between $P(A)$ and $Q_k(A)$ for various A (or the minimal logarithm of the relevant quotient, etc.) is ϵ, then we could require, e.g., that the maximal probability of E_k in the extended space should be $1 - \epsilon$? There are obviously many variants which can be explored, but it is not trivial that in each case a suitable extension algorithm will exist.

Second, is it possible to produce a similar extension algorithm which satisfies the additional requirement that according to the agent "some expert is right", namely, that $\bigcup_{1 \leqslant k \leqslant n} E_k = W'$? This might be especially interesting in the context of variants of the Principal Principle, where expert functions would be possible objective chance functions, and the disjunction of all considered propositions of the form "The objective chance functions is given by *ch*" should be a tautology.

Third, the construction offered here cannot (outright) model nesting, that is, having opinions regarding experts who have their own opinions regarding experts, etc.

Fourth, it might be practical to have an algorithm which would take as additional input the agent's degree of belief that a particular expert is correct, that is, the target $P'(E_k)$ for one or more k. It is not evident for which sets of such inputs constructing an appropriate extension is possible.

Producing an algorithm which would achieve one of the above goals is another task for the future.

Part II

Measuring the value of one's credal state

Chapter 5

A case for Inverse Relative Entropy

Suppose that an agent at a certain time has credences in propositions which form a Boolean algebra. These credences, or degrees of belief (we will use these terms interchangeably), are given by a "belief function", $\mathfrak{b}$. Suppose that subsequently the agent obtains evidence which entails constraints that $\mathfrak{b}$ does not meet. Which belief function should the agent adopt? This chapter is concerned with problems of this kind ("belief update problems"), in particular with two specific issues posed as open problems at the end of Leitgeb and Pettigrew (2010b). The larger context is assessment of the qualities of different ways of approaching belief update problems in general; we will see a few disadvantages of some proposed solutions, and as you may guess from the title of this chapter, I would like to suggest that the method involving the minimization of inverse relative entropy is a promising one.

Some belief update problems have answers with which many people feel comfortable; the most common example would be that if, for some proposition E, the agent learns that E (and nothing more), she should update her belief function by conditionalization, provided that her initial credence in E was not equal to zero (more on this in Section 5.1). However, with respect to other Problems, intuitions may vary wildly. What if the evidence entails (only) new credences in two overlapping propositions? (This is the "simultaneous update problem".) Or what if it entails (only) a new conditional credence? (This is the "Judy Benjamin problem".) Some might want to say that in such cases there is no unique answer; that is, that the evidence underdetermines the choice of the new belief function. Still, some answers may be more reasonable than others. Given a belief update framework, it may be fruitful to check whether it suggests a specific reaction in such cases, and if so, whether its form may suggest some shortcomings of the framework we might want to try to avoid.

What follows requires some setting up. We assume that we are considering an agent who holds credences in a Boolean algebra of propositions which are subsets of a finite set, W, which has n elements.[1] We can think of W as containing worlds epistemically possible for the agent. Each $w_i \in W$ is to be identified with a unit vector in $\mathbb{R}^n$ (with 0s on all axes apart from the ith and a 1 on the ith). The credences are given by a function b. We assume we are dealing with probabilistic agents—that is, b is a classical probability function. Thanks to this assumption, we can think of a belief function b as an element of $\mathbb{R}^n$, namely, a tuple containing the values of $b(\{w_i\})$ for all i. Throughout the chapter, b, if not treated as a variable, will always be the "prior" belief function of the considered agent, that is, the one she has before she receives the evidence which is the topic of the given belief update problem, and b' will be the function to be adopted after she receives that evidence. Therefore, b will be the "prior" belief function, and b' the "posterior" one.

The structure of the chapter is as follows. In Section 5.1 I propose a distinction between belief update *methods* and *rules* which I hope will be beneficial to the clarity of the arguments. Section 5.2 covers the Judy Benjamin problem. Subsequently, in Section 5.3 I tackle the simultaneous update problem. For ease of reading, the proofs of the results given there are relegated to the Appendix (p. 135). Then, in Section 5.5, I introduce a new, "symmetric" variant of the Judy Benjamin problem and use it to present other features of the discussed belief update methods which differentiate them. I finish the chapter by writing out the proof that the inverse relative entropy is a Bregman divergence

---

[1] We make the finitude assumption following Leitgeb and Pettigrew (2010a).

and briefly discussing the relationship between the different divergence-style formulas used in various corners of the literature.

5.1. Belief update methods and rules

One can propose a way of updating belief functions by formulating statements which may be varyingly specific. One can, for example, say "always, if possible, minimize relative entropy". This way, one has a pleasingly general proposal which may, on the other hand, be at first glance uninformative in specific cases: how does one go about obtaining the needed belief function in a given situation? On the other hand, one can say, for example, "when you learn that E is true (and nothing more), adopt, if possible, the belief function b' such that, for any proposition A, $b'(A) = b(A|E)$". This way, one has an easy rule for constructing the new belief function provided the evidence is in a certain format. I will call statements of the first kind "methods", and those of the second kind "rules". Methods do not distinguish between the different types of constraints which are implied by evidence, but rules do. Rules can be motivated by methods; in fact, once the appropriate definitions are given, one may attempt to prove theorems to the effect that a rule (say, Conditionalization—see below, p. 102) follows from a method (say, Minimizing Relative Entropy—see below, p. 101). If there are convincing arguments against a rule which follows from a given method, then they speak also against that method. In this section I will give definitions of methods and rules which I will use later.

5.1.1. Methods

Suppose one subscribes to the popular project of accuracy-first epistemology[2] and claims to have a specific inaccuracy measure $I(b, w_i)$ which associates with each belief function b and possible world w_i a number which, intuitively, is to be a measure of b's "distance from truth" if w_i is the actual world. Then one could propose a belief update

---

[2] For a recent monograph see Pettigrew (2016). For the field-opening paper see Joyce (1998).

method by saying "always, if possible, minimize the expected inaccuracy". The specifics will depend on the inaccuracy measure proposed and the chosen way of calculating its expected value. In the case of the framework from Leitgeb and Pettigrew (2010a), which we will henceforth dub the "LP" framework, and in the context of which the two problems mentioned above will first be considered, so-called "quadratic inaccuracy measures" are used. The relevant method is the following:

The Quadratic Update Method (QUM):
Given evidential constraints C, the belief function b' which should be adopted is such that it satisfies C and minimizes the expression

$$\sum_n b(\{w_n\})\|w_n - b'\|^2.$$

($\|\cdot\|$ is the Euclidean norm.) Remember that b is the agent's belief function before she receives new evidential constraints, and that since we assume that belief functions are probability functions, b' can be considered as a vector. The "the" in the formulation of (QUM) is there because for all update problems discussed in this chapter the empirical constraints define a convex and closed set of belief functions, so the appropriate minimum exists. This is the case for all update methods considered here.

After giving a critique of the LP framework, Levinstein (2012) proposes that the inaccuracy of a function b if w_i is the actual world be captured just by $-\ln b(\{w_i\})$.[3] This leads to the following method:

The Local-logarithmic Update Method (LLM):
Given evidential constraints C, the belief function b' which should be adopted is such that it satisfies C and minimizes the expression

$$-\sum_n b(\{w_n\}) \cdot \ln b'(\{w_n\}).$$

[3] Those familiar with the formal epistemology literature should recognize this as using "the only local proper scoring rule", see, e.g., Dawid, Lauritzen and Parry (2012). We will return to this topic in the next chapter.

Notice that both of these methods pay attention to the expected inaccuracy of a considered future belief function as calculated from the perspective of the current one, that is, one which the agent finds to be incompatible with the evidence. For doubts regarding the reasonability of this see, e.g., Pettigrew (2016, Chapter 15). If I were to mount a defense of this, I would attempt to say that it is only due to having the particular initial function that what the agent experiences is interpreted as this, and not some other, piece of evidence, for example learning that a particular proposition E is true. Spelling this out properly would require a whole formal system that models evidence, but I am unable to provide one at this point.

The remaining two methods I would like to introduce here use the notion of relative entropy. The first one is widely used in epistemology, the second one less so.

The Method of Minimizing Relative Entropy (MRE):
Given evidential constraints C, the belief function b' which should be adopted is such that it satisfies C and minimizes the expression

$$\sum_n b'(\{w_n\}) \ln \left(\frac{b'(\{w_n\})}{b(\{w_n\})}\right).$$

The Method of Minimizing Inverse Relative Entropy (MIRE):
Given evidential constraints C, the belief function b' which should be adopted is such that it satisfies C and minimizes the expression

$$\sum_n b(\{w_n\}) \ln \left(\frac{b(\{w_n\})}{b'(\{w_n\})}\right).$$

Notice that the only difference between the two expressions to be minimized is the position of the variable. In the MRE expression the variable, that is, the value of the "new" belief function b', occurs twice in each summand, while in the MIRE expression it does so only once since b and b' switch their roles. Readers who have encountered these expressions before might wonder about the choice of the labels; they might for example have arguments for the reverse labelling of the notions, referring to the first one above, and not the second, as minimizing *inverse* relative entropy. I do not want to enter into such debates; if one wants to label these notions in a different way I will not oppose. I would not like to suggest any particular "directionality" here, so I will avoid writing that, say, MRE minimizes relative entropy "from" b to b', or vice versa. For obvious reasons of (lack of) symmetry, I will also never write that what we are minimizing is the (inverse) relative entropy "between" b and b'. I believe that doing either of these things can only cause confusion.[4] I am using the labels as they appear in the recent and influential paper Douven and Romeijn (2011). In the 1981 paper in which van Fraassen introduced the Judy Benjamin problem, he referred to the MRE method as "InfoMin", that of minimizing relative information. (I will return to these issues in Section 5.6 below.)

The first small observation I would like to make in this chapter is the following: **the method of minimizing inverse relative entropy is equivalent to the local-logarithmic method**. Simply, the relevant expressions share their minima. The MIRE expression looks similar to an expected value calculated from the perspective of the belief function b. This might be interesting for those who would like to link entropy with epistemic inaccuracy. I will not pursue this matter here.

5.1.2. Rules

Here I will introduce the belief update rules—the ways of updating one's belief function in response to evidence of a specific type—which we will use later. The first one is the well-known conditionalization rule.

The Full Conditionalization Rule (FC):
Evidence: entails (only) that a proposition E is true.
Response: if $b(E) > 0$, then b' should be such that for any proposition A, $b'(A) = b(A|E)$.

---

[4] For an example, see the "Talk" page of the Wikipedia entry for the Kullback-Leibler divergence, Section 12.

It can be shown by a limit argument that the FC rule follows from both MRE and MIRE (see remarks in Chapter 15 of Pettigrew, 2016). How this relates to QUM is a more delicate matter to which we will turn soon. In what follows I will refer to the evidence which features in FC—namely, evidence which entails (only) that a proposition E is true—by the name "Bayesian evidence".

Let us now present the rule which many think to be a generalization of FC:

The Jeffrey Conditionalization Rule (JC):
Evidence: entails (only) new credences $q_1, \ldots, q_m$ in propositions $E_1, \ldots, E_m$ which form a partition of W.
Response: if for any $i \in \{1, \ldots, m\}$ $b(E_i) > 0$, then b' should be such that for any proposition A

$$b'(A) = \sum_{i=1}^{m} q_i \cdot b(A|E_i).$$

That the JC rule follows from both MRE and MIRE methods can again be shown by a limit argument (see Pettigrew, 2016). This was proven in Diaconis and Zabell (1982) for cases in which every q_i is nonzero.

However, as shown in Leitgeb and Pettigrew (2010b), the quadratic method used in the LP framework entails a different response to the same information:

The Alternative Jeffrey Conditionalization Rule (AJC):
Evidence: entails (only) new credences $q_1, \ldots, q_m$ in propositions $E_1, \ldots, E_m$ which form a partition of W.
Response: For any $i \in \{1, \ldots, m\}$ let d_i be the unique number such that

$$\left(\sum_{\{w \in E_i : b(\{w\}) + d_i > 0\}} b(\{w\}) + d_i \right) = q_i.$$

Then b' should be such that for any $w \in W$ such that $w \in E_i$,

$$b'(\{w\}) = \begin{cases} b(\{w\}) + d_i & \text{if } b(\{w\}) > 0 \\ 0; & \text{otherwise.} \end{cases}$$

In what follows, I will refer to the evidence which features in JC and AJC—namely, evidence which entails (only) new credences in propositions forming a partition of W—by the name "Jeffrey evidence".

Since we assume probabilism, AJC uniquely determines b'. Notice that AJC is applicable in some cases in which JC is not since the former does not need to assume that for each proposition E_i the initial credence in it is nonzero. (Consult Leitgeb and Pettigrew (2010b, p. 255) for a geometric interpretation of AJC.) The rule has been criticized by Levinstein (2012) for its various shortcomings; among those not covered in that paper but discussed by Leitgeb and Pettigrew themselves is AJC's cardinality dependence: unlike in the case of JC, the new credence in an arbitrary proposition A depends not only on the prior credences in the various E_i's and the conditional credences of the form $b(A|E_i)$, but also on how many elements the E_i's have.

You might justifiably think that AJC is incompatible with FC. It is easy to construct a W, a belief function b and a proposition E such that if we first assume that one's evidence entails that $b'(E) = 1$ and $b'(\neg E) = 0$ and then use AJC, we will end up with a different function b' from the one we would have reached had we initially assumed that the evidence entailed that one learned that E and then used FC. However, in the LP framework both AJC and FC are valid. This is because the authors think that the two situations should be modelled differently. Namely, the authors think that learning that E requires one to narrow one's set of epistemically possible worlds while obtaining credence 1 in E and credence 0 in $\neg E$ does not. This is just one of the subtleties which lead them to endorse something different from the QUM above in cases when one learns that some proposition is true.[5] In this way they are able to obtain the surprising result that their framework endorses both AJC

[5] Another subtlety concerns the distinction between "local" and "global" inaccuracy, which we will discuss in Chapter 6; see Leitgeb and Pettigrew (2010a, p. 204).

and FC. Fortunately, for all other belief update situations concerned in Leitgeb and Pettigrew (2010b), apart from "learning that E", and in particular for the situations which are at the heart of the two problems to which we will now turn, the authors decide on QUM as the belief update method.[6]

I will now turn to the two belief update problems posed in Leitgeb and Pettigrew (2010b, p. 262–263), adjusting the wording to the formalism used in the current chapter. I will approach the problems in what I believe to be the order of increasing difficulty.

5.2. The Judy Benjamin problem

The essence of the problem is the following: for some propositions A and B, your evidence entails (only) your new conditional credence in A given B. What should your new belief function look like?

Observe that JC cannot be used, at least not until the evidence is processed in some way which would lead to establishing new credences in all elements of some partition of W. We need, then, a different rule. Some people think there are many belief functions it would be rational to adopt in this situation; in other words, that no unique solution exists. A recent detailed exposition of such a view can be found in Huisman (2014). A (too) quick response would be that formal epistemology treats belief update problems in the context of *ideal* rationality, which may imply in every evidential situation a unique rational belief update procedure, even if we, real agents, can only attempt to imperfectly imitate this ideal. Another direction would be to notice that there are obviously ridiculous ways of responding to such evidence (say, "set $b'(B)$ to 1"), and some other responses which many people find to be unintuitive (more on that below). If one's update method prescribes some such response, this would be to its disadvantage.

Here is the original illustration of the problem from the paper as introduced by van Fraassen (1981):

---

[6] The authors call it "minimizing expected global inaccuracy"; as before, see Chapter 6 below.

> [Judy Benjamin (a soldier)] and her patrol are dropped in a swampy area which they have to patrol. The area is divided into the region of the Blue Army, to which Judy Benjamin and her fellow soldiers belong, and that of the Red Army. The Red Army region is further divided into Headquarters Company Area and Second Company Area. The patrol has a map which none of them understands, and they are soon hopelessly lost. Using their radio they are at one point able to contact their own headquarters. After describing whatever they remember of their movements, they are told by the duty officer 'I don't know whether or not you have strayed into Red Army territory. But if you have, the probability is 3/4 that you are in their Headquarters Company Area.' At this point the radio gives out. (van Fraassen (1981))

There seem to be three propositions of interest: B ("JB's unit is in the Blue Army region"), R_1 ("JB's unit is in the Red Army region, Headquarters Company Area"), and R_2 ("...Red Army region, Second Company Area"). B, R_1 and R_2 are pairwise incompatible and jointly exhaustive; that is, they form a partition of whatever W we use as the set of Judy's epistemically possible worlds. The evidence Judy receives seems to entail (only) that $\mathsf{b}'(\mathsf{R}_1) = 3 \cdot \mathsf{b}'(\mathsf{R}_2)$.[7] Her prior belief function b is set to $\mathsf{b}(\mathsf{R}_1) = \mathsf{b}(\mathsf{R}_2) = 0.25$, $\mathsf{b}(\mathsf{B}) = 0.5$. Van Fraassen's (1981) discovery is that according to MRE the belief function b' which Judy should adopt is such that $\mathsf{b}'(\mathsf{B}) = 0.532$, approximately. (This holds regardless of the cardinality of W.) Many, including van Fraassen himself, found this result to be unintuitive: why should Judy's degree of belief that her unit is in the Blue Army region increase? What the duty officer says seems to concern only the Red Army areas. In fact, for any fraction different from 3/4 (and 1 or 0) the situation is similar: MRE has it that Judy's credence in B should increase. (See Lukits (2014) for a recent defense of this view.) I will now argue that the LP framework shares this unintuitive feature because it employs QUM as its belief update method.

I have already mentioned above that the updating rule prescribed by the LP framework in cases involving Jeffrey evidence displays de-

---

[7] In this presentation we abstract away from details like the need to model Judy's beliefs in the reliability of her duty officer. For an approach exploring this direction, see Grove and Halpern (1997).

pendence on cardinalities. It turns out that the same is true for the Judy Benjamin problem. Let us first consider the smallest possible case, with W as $\{w_1, w_2, w_3\}$, $R_1 = \{w_1\}$, $R_2 = \{w_2\}$, $B = \{w_3\}$, and with the initial belief function b as the triple $\langle 0.25, 0.25, 0.5 \rangle$. A short exercise in minimizing a real-valued function of one variable shows that according to QUM the belief function b' that Judy should adopt is such that $b'(B) = 7/13$. So, just like with MRE, the LP framework will have Judy increase her degree of belief that her unit is in the Blue Army region. Still keeping with the 3-element W, we can generalize a bit with respect to the evidence given and assume that the evidence entails, for some non-negative m, that $b'(R_1) = m \cdot b'(R_2)$.[8] A slightly longer exercise in analysis[9] shows that if m is not equal to 1, then according to QUM, $b'(B) > b(B)$. Therefore, we can see that if we use the smallest possible (3-element) model for the situation, then QUM performs similarly to MRE, that is, it shares with it the disadvantage of not giving the intuitive solution (which would keep the degree of belief in B constant). I do not think the particular numeric details of the posterior credences are important.

A variant of the Judy Benjamin problem involves splitting the Blue Army region into two areas, mirroring those of the Red Army region, and assuming that Judy's initial belief function is uniform with regard to the placement of her unit in the four regions involved. Consider, then, the case in which $W = \{w_1, w_2, w_3, w_4\}$, $b(\{w_i\}) = 1/4$ for any $i \in \{1, \ldots, 4\}$, $R_1 = \{w_1\}$, $R_2 = \{w_2\}$, $B_1 = \{w_3\}$, $B_2 = \{w_4\}$, and $B = B_1 \cup B_2$. If, as above, the evidence entails (only) that $b'(R_1) = 3 \cdot b'(R_2)$, then QUM dictates that $b'(B) = 5/9$. Therefore, according to the LP framework Judy's posterior credence in B depends on the cardinality of the propositions involved.[10] I do not want to claim that this is a particular problem for the framework since I believe the points from

---

[8] Observe that this is equivalent, since R_1 and R_2 are disjoint, to the evidence entailing (only) a specific new conditional credence. In the original example Judy learns that $b'(R_1 | R_1 \vee R_2) = 3/4$, which (if $b'(R_1)$ and $b'(R_2)$ are not equal to 0) is equivalent to $b'(R_1) = 3 \cdot b'(R_2)$.

[9] I omit the arguments for this and the previous observation. They are similar to the proofs of Facts 8 and 9 below given in the Appendix (p. 135).

[10] I am unable to give a general account of this cardinality dependence. For the current purposes it is enough to notice that the dependence exists.

the previous paragraph already show that its belief update method, QUM, does not fare better than MRE in response to the Judy Benjamin problem.

Consider, however, what LLM would mandate as the solution here. Take the original formulation of the problem and start with the simplest case, with W having 3 elements. The prior belief function b is, again, $\langle 0.25, 0.25, 0.5\rangle$. The posterior belief function b' can be written as $\langle 3\beta, \beta, 1-4\beta\rangle$. The expression we should minimize according to LLM (p. 100) is thus just a function of β and is equal to the following:

$$-0.25 \cdot \ln(3\beta) - 0.25 \cdot \ln\beta - 0.5 \cdot \ln(1-4\beta) \tag{5.1}$$

the derivative of which is

$$-\frac{0.25}{\beta} - \frac{0.25}{\beta} + \frac{2}{1-4\beta}$$

which equals 0 exactly when

$$\beta = 0.125$$

where (since the second derivatives are positive) the expression (5.1) reaches its minimum. Therefore, the belief function minimizing the expected inaccuracy as given by the logarithmic measure is $\langle 0.375, 0.125, 0.5\rangle$, which means that $\mathrm{b}'(\mathrm{B}) = \mathrm{b}(\mathrm{B}) = 1/2$. A simple Lagrangian Multiplier argument extends this to cases of arbitrary finite cardinality of all propositions involved.

This is not a novel finding, but an easy (I think) way of showing a result by Douven and Romeijn (2011), who found that the MIRE method does give the "intuitive" solution to the Judy Benjamin problem. Since we have already observed that LLM is equivalent to MIRE, the "good" behavior of LLM shown above should not be surprising.

So far we have seen that the LP framework's answer to the Judy Benjamin problem is "unintuitive" in the same sense in which MRE's answer is: the degree of belief in a proposition which is not explicitly talked about is supposed to increase. In addition, the answers mandated by the LP framework are cardinality dependent, as opposed to those given by MRE. MIRE offers both the intuitive and cardinality independent solution to the problem. It also has the pleasing feature of giving a graspable reduction of the Judy Benjamin problem to one solvable by Jeffrey Conditionalization—"graspable" in the sense that it can be given only by referring to old credences and new constraints, as we will see shortly. In contrast, any application of MRE to the Judy Benjamin problem with a particular m arrives at a belief function which is also obtainable using Jeffrey Conditionalization, but no "graspable" reduction seems to exist.

The rule which MIRE endorses for the Judy Benjamin problem:
Evidence: entails (only), for some disjoint propositions A and B and a nonzero positive real number m, that $b'(A) = m \cdot b'(B)$.
Response: Consider this as a case in which the evidence entails (only) new credences in all elements of the partition $\{A, B, \neg(A \vee B)\}$:

- $b'(\neg(A \vee B)) = b(\neg(A \vee B))$;
- $b'(B) = \frac{b(A \vee B)}{m+1}$;
- $b'(A) = 1 - b'(\neg(A \vee B)) - b'(B)$

and use the JC rule.

If we generalize the evidential situation so that it involves an arbitrary number of finite disjoint propositions, we will see that MIRE endorses a rule called "Adams Conditionalization", which is a special case of JC (see Bradley, 2005).

In what follows, I will refer to the evidence which features in the Judy Benjamin problem—namely, evidence which entails (only) a new conditional credence, or equivalently a proportion of credences in two disjoint propositions—by the name "Judy Benjamin evidence".

5.3. The Simultaneous Update problem

We will now turn to the second belief update problem to be discussed. It boils down to the following: what is one's rational posterior belief function if one's evidence entails (only) one's new credences in *overlapping* propositions? (I will call such evidence "overlapping evidence".)

That is, suppose you know of your new belief function b' that $b'(A) = p$ and $b'(B) = q$ for some p and q, but as a matter of fact $A \cap B \neq \emptyset$. What is the precise shape your b' should take?

Notice that the situation does not allow us to use JC (or AJC) right away: we are not given new credences in all elements of some partition. What we do receive are new credences in all elements of *two* partitions: $\{A, \neg A\}$ and $\{B, \neg B\}$. One direction in which we could proceed would be to argue for some transformation of the data entailed by the evidence so that new credences in all elements of a single partition would be obtained; seemingly the most natural partition to be used here consists of the four logical combinations of A and B. If this were successful, we could then use an appropriate update rule, for example JC or AJC, that is compatible with our chosen belief update method. I will argue below that in the case of the LP framework, such an approach is only partially successful, while the goal is obtained with no reservations if we use MIRE.

The problem is called a "simultaneous update" because it relates to a well-known feature of Jeffrey Conditionalization (shared by AJC): its noncommutativity. JC is sensitive to the order in which it is applied to numerical constraints following from the evidence. That is, applying JC twice, using two sets of new credences in elements of some two partitions, will typically lead to a different belief function depending on which set is used first. Some people consider this to be a problem since the new credences are taken to correspond to the evidence received and the order of evidence should—they say—not matter for the eventual belief function. In my opinion this worry is insubstantial because changing the order of using the numerical constraints usually has to change the evidence from which these constraints are supposed to follow. (Think about the different evidence needed to change your degree of belief that the coin you are repeatedly tossing is fair to 0.9 depending on whether your initial credence in that is 0.01 or 0.8.) This is not an original view; Lange (2000) and Osherson (2002) present arguments to the effect that changing the order of application of the purely numerical constraints results in describing two different evidential situations.[11] Still, there

[11] This does not put the issue fully to rest. See Weisberg (2009) on "commutativity on experiences".

have been attempts at combining the data about two partitions into one, thus reducing two updates to be performed consecutively into a single "simultaneous" one. An example of this is the recent paper by Park (2013), in which the author analyses in detail an earlier suggestion from Williams (1980) regarding the application of MRE to that task. The subsection 5.3.3 below can be seen as complementing the results of Park (2013), using MIRE instead.

5.3.1. How QUM approaches the problem

We will first consider the simplest non-trivial case, which involves four possible worlds and two two-world propositions sharing one world. Notice that in this case the task reduces to figuring out $b'(A \cap B)$. It turns out that the LP framework, with its choice of QUM as its belief update method, offers a simple solution to the simultaneous update problem in this case (see the Appendix for the proof, p. 137, but remember the note on notation just before the argument):

Fact 8. *Let $W = \{w_1, w_2, w_3, w_4\}$ be the set of epistemically possible worlds for some agent. Let A be the proposition $\{w_1, w_2\}$; let B be $\{w_2, w_3\}$. Suppose that the agent's evidential constraints are that her posterior belief function b' satisfies the following: $b'(A) = p$ and $b'(B) = q$, for some p and q. Let us label the number $b(A \cap B) + \frac{p - b(A) + q - b(B)}{2}$ as K. The belief update function which the agent should adopt according to QUM is fully determined by the two constraints and the following condition:*

$$b'(A \cap B) = \begin{cases} 0 & \text{if } K < 0 \\ \min\{p, q\} & \text{if } \min\{p, q\} < K \\ K & \text{otherwise.} \end{cases}$$

That is, the change in your credence in $A \cap B$ is the average of the changes in your credences in A and in B (whenever it makes mathematical sense, i.e., you don't go above 1 or below 0). I would say that the solution is at least not immediately absurd. I will now illustrate it by an example using a modified story from Osherson (2002).

Suppose that listening to the radio I hear a forecast for rain but I'm not sure whether it comes from the chief meteorologist or from his unreliable deputy. After the broadcast concludes I have the following belief function b_1, where R is "It rains today" and C is "The chief was speaking":

	$\neg RC$	RC	$R\neg C$	$\neg R\neg C$
b_1	.2	.4	.1	.3

Now, a glance at the sky raises my credence in R to .7. Suppose also that the forecast is rebroadcasted and even though I strain my ears, I conclude I should not change my credence in C. The following is my new credence as dictated by Fact 8:

	$\neg RC$	RC	$R\neg C$	$\neg R\neg C$
b_1'	.1	.5	.2	.2

This is because my credence in R increases by .2, and my credence in C stays the way it was. The average of these changes is .1, which according to Fact 8 is the increase in my credence in RC. In this simple case, all other credences are directly calculable.

Even though the solution of the problem offered by QUM in the simplest non-trivial case may look reasonable, I would like to raise some problems regarding it:

1. I will show that the simple method offered by Fact 8 does not work if the cardinality of the events is different; that is, in the LP framework the information entailed by the agent's evidence is not directly translatable into information about a partition (Section 5.3.1.1);

2. I will also argue that in the four-world case the updating rule dictated by the LP framework already leads to unfortunate updating behavior (Section 5.3.1.2).

5.3.1.1. Cardinality dependence again

Given Fact 8, at this point a *prima facie* reasonable way to proceed inside the LP framework would be the following:

- given any similar situation with an arbitrary finite W, calculate $b'(A \cap B)$ using the above formula;

- this, together with the two constraints, would presumably give us the posterior credences in all propositions from the set $\{A \cap B, A \cap \neg B, \neg A \cap B, \neg A \cap \neg B\}$, which is a partition of W; in this way we would obtain constraints such as would follow from Jeffrey evidence;
- having that, we should now (it would seem) use AJC to derive the full shape of b' (since we know this is the rule QUM leads to in response to Jeffrey evidence).

It turns out, however, that the LP framework itself would judge this procedure as wrong. It turns out the answer **depends on the cardinality of the propositions**. We already knew that cardinality is a factor that plays an important role for AJC; we will now see that in situations involving new credences in two overlapping propositions cardinality is relevant also for the **input** for AJC; depending on the cardinality of the propositions involved, the new credences in the four logical combinations of the two propositions will be different. I think that this is an unfortunate consequence which we should try to avoid. I will now illustrate it by the following Fact and a modification of the previously used weather forecasting example. The relevant feature of the LP framework is the following (for the proof and some discussion of the mathematical details, see again the Appendix, p. 137, but remember the note on notation on p. 135):

Fact 9. *Let* $W = \{w_0, w_1, w_2, w_3, w_4\}$ *be the set of epistemically possible worlds for some agent. Let* A *be the proposition* $\{w_0, w_1, w_2\}$*; let* B *be* $\{w_2, w_3\}$*. Suppose that the agent's evidential constraints are that her posterior belief function* b' *satisfies the following:* $b'(A) = p = b(A)$ *and* $b'(B) = q$*, for some* p *and* q*. If the belief update function which the agent adopts is the one fully determined by the two constraints and the following condition*

$$\begin{aligned} b'(A \cap B) &= b(A \cap B) + 4/7(q - b(B)), \\ b'(\{w_0\}) &= b(\{w_0\}) - 2/7(q - b(B)), \\ b'(\{w_1\}) &= b(\{w_1\}) - 2/7(q - b(B)), \end{aligned}$$

then it is the one which QUM dictates the agent should adopt in this case.

I do not want to suggest that this is in any way a profound discovery: it is a result of a simple minimization argument which will allow me to produce examples which point, in my opinion, to some deficiencies of QUM.

Consider, then, a modification of the previous story about the chief meteorologist and the weather. The goal is now to split one possible world into two worlds—to transform the previous four-world case to an example where the proposition ¬RC contains two worlds and so the whole space contains five worlds. Suppose, then, that I owe the chief meteorologist money and I don't want to meet him. I know that whenever it doesn't rain, he always walks home through a park he never otherwise visits and in which I walk my dog (which I do regardless of the weather); not wanting to disturb my dog's routine, I will also go to the park today. (If it rains, the chief meteorologist takes a bus home.) The chief meteorologist can traverse the park via one of two paths, call them "1" and "2". Let B be the proposition "the chief meteorologist will walk via path 1 today". My four epistemic possible worlds

$$\neg RC \quad RC \quad R\neg C \quad \neg R\neg C$$

become

$$\neg RCB \quad \neg RC\neg B \quad RC\neg B \quad R\neg C\neg B \quad \neg R\neg C\neg B$$

and my initial credence becomes, say,

	¬RCB	¬RC¬B	RC¬B	R¬C¬B	¬R¬C¬B
b_2	.1	.1	.4	.1	.3

(it is not important that my credences in ¬RCB and ¬RC¬B are equal: they just need to sum up to .2). Notice that the ¬RC world has effectively split in two; think of the two stories as describing two agents with a similar but different space of epistemically possible worlds.

Consider now the same evidential situation and suppose (which I think is intuitive enough[12]) that it gives rise to the same numerical

[12] Although I realize that for some it would need an additional argument regarding context (in)sensitivity and related issues. I hope to tackle these topics in a future paper.

constraints: a glance at the sky raises my credence in R to .7; the forecast is rebroadcasted and even though I strain my ears I conclude I should not change my credence in C. The following is, by Fact 9, my new credence, as mandated by QUM:

	$\neg RCB$	$\neg RC\neg B$	$RC\neg B$	$R\neg C\neg B$	$\neg R\neg C\neg B$
b_2'	.043	.043	.514	.186	.214

Now recall the initial and updated credences for the first agent:

	$\neg RC$	RC	$R\neg C$	$\neg R\neg C$
b_1	.2	.4	.1	.3
b_1'	.1	.5	.2	.2

Comparing b_1' and b_2' we notice that the two agents:

- started with the same credences in R, C, and RC[13];
- faced the same evidence implying (only) that they should increase their credence in R by the same amount and not change their credence in C;
- ended with a different credence in RC;
- and it seems the only difference was whether the park was considered or not.

In my opinion, this updating behavior is unintuitive: the role of the park is unclear at best. It would seem to me that we should avoid this kind of cardinality dependence if we can; in the next section I will show that the MIRE method succeeds in this. I also do not see a way of invoking context-dependence, which for some is a natural way of replying to similar cardinality-dependence-related issues troubling Leitgeb's (2014) stability theory.

---

[13] This is not literally true, for example since C is for one agent a doubleton, and for the other a three-element set. To be technically correct I would have to write something like "the agents started with the same credences in the propositions (sets of worlds) they associate with the sentence 'the chief was speaking' "; I decided to avoid the verbiage.

There are also cases in which not only the credence in a suitably chosen proposition ends up being different, but the two agents end up with different probability *rankings* regarding two propositions which they initially believe to the same degree (and so rank in the same way); that is, cases in which for some two propositions ϕ and ψ

$$b_1(\phi) = b_2(\phi) \text{ and } b_1(\psi) = b_2(\psi)$$

but

$$b_2'(\phi) < b_2'(\psi) \text{ while } b_1'(\phi) > b_1'(\psi).$$

As an example, consider two agents with sets of epistemically possible worlds and propositions as discussed above, and suppose the initial belief functions of the agents look like this:

	¬RC	RC	R¬C	¬R¬C
b_1	.21	.3	.3	.19

	¬RCB	¬RC¬B	RC¬B	R¬C¬B	¬R¬C¬B
b_2	.105	.105	.3	.3	.19

Notice that $b_1(\neg RC) = .21 = b_2(\neg RC)$ and $b_1(\neg R\neg C) = .19 = = b_2(\neg R\neg C)$. So, both agents initially rank ¬RC as more probable than ¬R¬C. Suppose now that the empirical constraints dictate that the credence in C is to stay the same, while the credence in R is to become .9. The following tables show the posterior belief functions of the two agents as calculated using Facts 8 and 9:

	¬RC	RC	R¬C	¬R¬C
b_1'	.06	.45	.45	.04

	¬RCB	¬RC¬B	RC¬B	R¬C¬B	¬R¬C¬B
b_2'	.019	.019	.472	.428	.062

Notice that $b_1'(\neg RC) = .06 > .04 = b_1'(\neg R\neg C)$, while $b_2'(\neg RC) = .038 < .062 = b_2'(\neg R\neg C)$. Therefore, despite the fact that the agents received the same empirical constraints regarding their new credences in R and C, one of them reversed her ranking of two logical combinations of these propositions.

5.3.1.2. QUM: The four-world cases

Cardinality issues aside, I will now argue that Fact 8 alone already has some unintuitive consequences which indicate deficiencies of the LP framework. Consider a different modification of our original story involving an agent with the initial belief function b_1 (p. 112). Suppose two pieces of evidence are given:

- a glance at the sky *lowers* the credence in R to .4 (by .1);
- a rebroadcasting of the forecast *increases* the credence in C to .7 (by .1).

The following is the new credence as dictated by Fact 8:

	$\neg$RC	RC	R$\neg$C	$\neg$R$\neg$C
b_1''	.3	.4	.0	.3

Because the average of the two changes is 0, the credence in the conjunction has to stay the same. Notice that, for unknown reasons, the agent has ended up with credence 0 in R$\neg$C—that is, the lowest possible for a probabilist. That by itself is unintuitive, but situations in which similar phenomena arise have already been described by Levinstein (2012). I want to add to this an interesting feature of the updating behavior which follows from Fact 8. What happened in the last example is that the agent updated her credences in R and C as if these two were *perfectly anticorrelated*; that is, she interpreted an increase in the credence in C as an increase in the credence in $\neg$RC, and she interpreted a decrease in the credence in R as a decrease in the credence in R$\neg$C. She acted as if she was either oblivious to the fact that she holds a positive credence in RC, or she thought the influence of the two pieces of evidence on her credence in RC cancels out exactly, which should be supported by some additional argument.

In general, we can notice that Fact 8 dictates the following update behavior:

- if increase in the credence in R equals the increase in the credence in C: update as if they are perfectly correlated (increase the credence in RC by the same amount);

- if increase in the credence in C equals the *negative* of the increase in the credence in R: update as if they are perfectly anticorrelated (leave the credence in RC as it was, increase the credence in ¬RC and decrease the credence in R¬C by the same amount),

whenever it makes mathematical sense (no credences end up being negative, etc.).

Even if I think that the main disadvantage of QUM presented in this section is the cardinality dependence (Subsection 5.3.1.1), the updating behavior just outlined may also be worrying: I would say this is not obvious and definitely requires some supporting argument. I will also note that this behavior extends to some extent beyond the four-world case, but I only calculated some specific five-world examples.

5.3.2. Cardinality dependence and fine-graining

Cardinality dependence is in my opinion a serious problem, not just because it sometimes is hard to explain, but more importantly because we might not know the cardinalities of the propositions we will be dealing with in some future phases of our belief update. That is, it might very well happen that some other features of the considered situation become apparent and relevant; this will "fine-grain" the description, effectively "dividing" the worlds. For a formal theory of this and concrete examples I suggest looking at the notion of "filtrations" (mentioned in the previous chapter) in the theory of stochastic processes (Nikeghbali (2006)) and how they are applied for example to insider training (Imkeller (2002)); for an application of similar ideas in the context of probabilistic causality, see Marczyk and Wroński (2014). *Any update method which, like QUM, features cardinality dependence, will not commute with fine-graining*. It would be safest in general, then, to use a belief update method which does not feature cardinality dependence. We will now see that using the MIRE method allows us to avoid the sort of cardinality dependence described here.

5.3.3. The MIRE solution to the problem

It turns out that the MIRE method (p. 101), or equivalently the local-logarithmic method, offers a solution to the simultaneous update problem which is independent of the cardinality of the propositions involved; that is, the problem is reduced to figuring out the credences in the four logical combinations of the propositions, at which point Jeffrey Conditionalization is invoked (for the proof, see again the Appendix, p. 139, and again remember the note on notation on p. 135).

Fact 10. *Let* A *and* B *be two propositions such that* $A \cap B \neq \emptyset$ *and let* b *be the agent's prior belief function. Suppose that the agent's evidential constraints entail (only) her new credences* $b'(A)$ *and* $b'(B)$*. To arrive at the belief function* b' *the agent should adopt according to MIRE, we should first calculate the agent's new credences in the logical combinations of* A *and* B *so that the following is true*

$$\frac{b(A \cap \neg B)}{b'(A \cap \neg B)} + \frac{b(\neg A \cap B)}{b'(\neg A \cap B)} = \frac{b(A \cap B)}{b'(A \cap B)} + \frac{b(\neg A \cap \neg B)}{b'(\neg A \cap \neg B)}$$

and then use Jeffrey Conditionalization.

If we write the new credences in A and B as p and q, we can rewrite the condition in Fact 10 as

$$\frac{b(A \cap \neg B)}{p - b'(A \cap B)} + \frac{b(\neg A \cap B)}{q - b'(A \cap B)} = \frac{b(A \cap B)}{b'(A \cap B)} + \frac{b(\neg A \cap \neg B)}{b'(A \cap B) + 1 - p - q}$$

at which point it is more immediately seen that what we have is just an equation with a single variable. Moreover, the solution—that is, the value of $b'(A \cap B)$—is always available for the agent (see the proof in the Appendix). I would say that an intuitive advantage of this solution—apart from what I take to be its modest mathematical elegance, though this of course is a matter of taste—is that the new credences in the four logical combinations of A and B are independent of the cardinalities of the propositions, and so no such unfortunate updating behavior as the one described in Section 5.3.1.1 can arise. This for me forms another argument in favor of using the MIRE update method instead of the quadratic one. Just to reiterate, in my opinion MIRE also has the edge over MRE because it delivers the "intuitive" solution to the Judy Benjamin problem.

The difference between the MIRE method and QUM is already visible in our original, four-world updating case: as the reader may verify, the eventual belief function $b'_{1_{MIRE}}$ dictated by the MIRE update method is approximately displayed in the following table:

	¬RC	RC	R¬C	¬R¬C
b_1	.2	.4	.1	.3
b'_1	.1	.5	.2	.2
$b'_{1_{MIRE}}$	0.107	0.493	0.207	0.193

The new credences allotted by MIRE to the four propositions will be as above, regardless of how many (but finitely many) worlds belong to each of them. The way of updating mandated by MIRE will also not exhibit the worrying features described on p. 117 ("updating as if the events were perfectly anticorrelated", etc.).

To connect this development with the results in Park (2013), I will now present a Fact containing a direct formula for the posterior belief function mandated by MIRE in response to the simultaneous update problem concerning some two propositions A and B. It will be convenient to use characteristic functions. Define, then, for any $w_i \in W$ and $E \in \{A, B\}$,

$$\chi_{E,i} = \begin{cases} 1 \text{ if } w_i \in E \\ 0 \text{ otherwise.} \end{cases}$$

Fact 11. *Let* A *and* B *be two propositions such that* $A \cap B \neq \emptyset$ *and let* b *be the agent's prior belief function. Suppose that the agent's evidential constraints entail (only) her new credences* $b'(A) = p$ *and* $b'(B) = q$*. The belief function* b' *the agent should adopt according to MIRE is defined as follows:*

For any $i \in \{1, \dots, n\}$, $b'(\{w_i\}) = \frac{b(\{w_i\})}{-\lambda_0 - \lambda_1\chi_{A,i} - \lambda_2\chi_{B,i}}$ *with parameters* λ_0, λ_1 *and* λ_2 *determined by the following three equations:*

$$\sum_{i=1}^{n} \frac{b(\{w_i\})}{-\lambda_0 - \lambda_1\chi_{A,i} - \lambda_2\chi_{B,i}} = 1;$$

$$\sum_{i=1}^{n} \chi_{A,i} \frac{b(\{w_i\})}{-\lambda_0 - \lambda_1\chi_{A,i} - \lambda_2\chi_{B,i}} = p;$$

$$\sum_{i=1}^{n} \chi_{B,i} \frac{b(\{w_i\})}{-\lambda_0 - \lambda_1 \chi_{A,i} - \lambda_2 \chi_{B,i}} = q.$$

For a proof, see the Appendix (p. 141). Unfortunately, neither Fact 10 nor 11 is easily generalizable either with respect to the number of partitions or the number of elements of the partitions.

One benefit of formulating the results in the above form is that it can immediately be seen that updating in the way mandated by MIRE is equivalent to updating using Jeffrey Conditionalization on the more "fine-grained" partition $\{A, \neg A\} \times \{B, \neg B\}$. One way to see this is to notice that if MIRE is used, then for any cell c of the more fine-grained partition there is a number r_c such that for any world w_i inside that cell $b'(\{w_i\}) = b(\{\omega_i\}) \cdot r_c$, which is characteristic of Jeffrey Conditionalization. After consulting the formulas in Park (2013, p. 3517) we see that the same is true for updating using MRE.

For those who would like to defend JC against the criticism of noncommutativity (p. 110) by presenting a way of reducing the two consecutive updates to one, there are then at least two options, one using MRE (see Park (2013)), the other using MIRE (presented here).[14] Which of them is better has to depend on the specifics of the given situation and on the interpretation of the relative entropy expressions the two methods employ. I leave investigating these issues for a future paper.

In this section I tried first to show what QUM says about the simultaneous update problem in certain types of cases, and next to argue that this points to some deficiencies of the LP framework. I then showed that MIRE provides a cardinality-independent solution which also allows adherents of Jeffrey Conditionalization to respond to some versions of the criticism regarding the rule's noncommutativity.

[14] Unfortunately, as mentioned before, in this chapter I only show the way MIRE deals with the issue if two partitions with two elements are involved. Park describes also the case of three two-element partitions.

5.4. (Intermediate) conclusions

The goal of this chapter was to respond to two open problems regarding belief update in the LP framework, which employs the quadratic update method. I have given partial answers in sections 5.2 and 5.3. I tried to argue that the answers exhibit some troubling features of QUM: most importantly, a sort of cardinality-dependence which has not, as far as I know, been discussed in the literature (Section 5.3.1.1), but also not giving the "intuitive" (p. 106) solution to the Judy Benjamin problem (Section 5.2). Along the way I tried to show that the method employing inverse relative entropy fares better in that regard.

The following table displays the status ("acceptable" / "not acceptable") of the three belief update methods' responses to the considered types of evidence from the perspective of someone who prefers the "intuitive" solution to the Judy Benjamin problem and cardinality-independent solutions to belief update problems in general:

	Jeffrey evidence	overlapping evidence	Judy Benjamin evidence
QUM	X	X	X
MRE	✓	✓	X
MIRE	✓	✓	✓

I do not want to suggest that MIRE trumps all other belief update methods, full stop. My goal was to argue that there are situations in which MIRE works at least as well as MRE, while QUM leads to unintuitive results. Of course, intuitions vary. Those, who—like van Fraassen—were dissatisfied with MRE's approach to the Judy Benjamin problem, and those who prefer cardinality independent solutions to belief update problems, should, I think, be relatively happy with what MIRE offers.

5.5. The symmetric Judy Benjamin Problem, or learning from conditionals whose antecedents form a partition

I have argued that both MRE and QUM do not answer well to the Judy Benjamin problem. We also already know that QUM suffers from cardinality-related problems that MRE and MIRE seem to be free of. In this section I will present a variant of the Judy Benjamin problem which will allow the display of some additional differences in the behavior of QUM and MRE. However, it is doubtful whether these differences point to some advantage one of these methods would have over the other.

Consider the four-world setup described on p. 107, with B_1 given the meaning "JB's unit is in the Blue Army region, Headquarters Company Area", and B_2 given the meaning "... Blue Army region, Second Company Area", and $R = R_1 \cup R_2$. Suppose the duty officer gives Judy the following information:

> I don't know whether or not you have strayed into Red Army territory. But if you have, the probability is 3/4 that you are in their Headquarters Company Area. Also, if you are still in the Blue Army territory—again, I don't know whether you are or not—the probability is just 1/3 that you are in the Blue Headquarters Company Area.

Note that in this case the information received can be thought of as containing two conditionals with exclusive and jointly exhaustive antecedents: loosely speaking, "if R" and "if B". That is, two posterior credences conditional on elements of a partition are imposed. We will call this a "symmetric" Judy Benjamin problem. It is symmetric in the sense that two posterior conditional degrees of belief are given, $b'(R_1|R)$ and $b'(B_1|B)$.

As can be seen in Figure 5.1, the situation does not differ dramatically from the original JB problem: MIRE tells us to keep our original credence in R (and so in B), while both MRE and QUM tell us to decrease it. It will be interesting to see how the situation depends on what conditional credences are specified in the added conditional regarding the Blue regions.

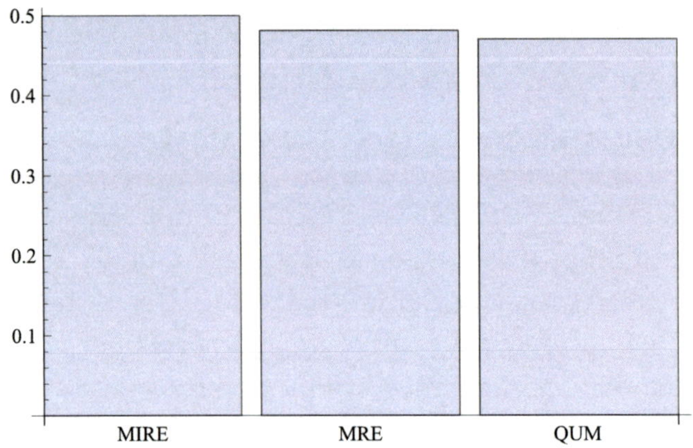

Figure 5.1. Posterior credence in R as mandated by the three update rules in which we are interested, in response to the information from the duty officer given in the quote on p. 123; this is a special case of the symmetric Judy Benjamin problem.

5.5.1. The symmetric Judy Benjamin problem, generalized

Consider, then a somewhat generalized form of the symmetric Judy Benjamin problem: the duty officer issues a statement which fixes the posterior credence $b'(R_1|R)$ at 3/4, and the posterior credence $b'(B_1|B)$ at some θ.

5.5.2. The uniform prior

Suppose the prior belief function is uniform over the four singletons. Figure 5.2 portrays the behavior of QUM, MIRE and MRE in this case.[15] The vertical axis is the posterior $b'(R)$. The horizontal lines mark the posterior $b'(R)$ in the original Judy Benjamin problem in its four-world variant—that is, when θ is not mentioned at all, and the prior is uniform—according to MIRE (red), MRE (blue) and QUM (green). The curves are graphs of the dependence of the posterior credence in R on the θ according to MRE (blue) and QUM (green), which both reach their minima for $\theta = 0.5$. That is, the answers given by both

[15] The figures in this chapter have been prepared using Mathematica. Please contact me for a notebook file if you are interested.

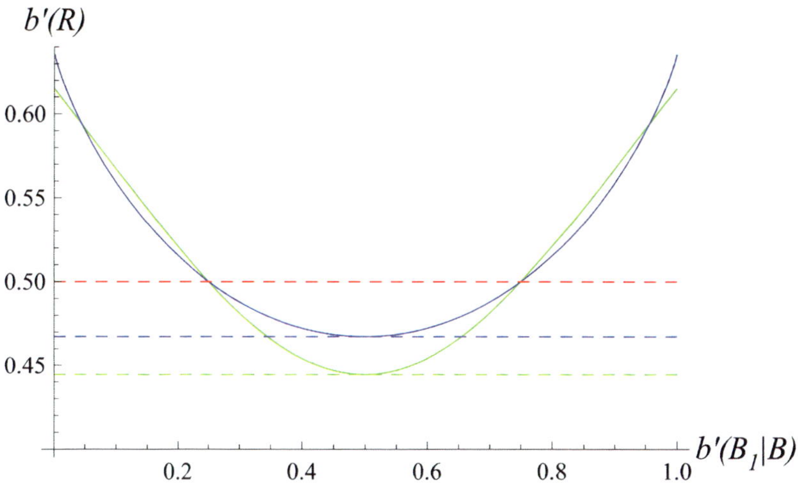

Figure 5.2. The symmetric JB problem for the uniform prior over four atoms.

QEM and MRE to the original JB problem mandate that the posterior $b'(B_1|B) = 0.5$—that is, this conditional probability is preserved.

Note that when $\theta = 3/4$, all three methods will not lead to any change in the credence in R: in a sense, the "influence" of one conditional balances the influence of the other.

Observe also that the two functions reach their minima on the corresponding horizontal lines. This means that, in the "uniform prior" case, according to both QUM and MRE one cannot specify a θ such that one would reach a posterior $b'(R)$ lower than the one mandated for the original JB problem. In other words, adding the information about the θ can only increase the recommended $b'(R)$ or keep it the same as in the original problem (in the very particular case of $\theta = 3/4$).

It is well known that for the original Judy Benjamin problem, MRE mandates keeping the prior uniformly divided between the singletons inside the B proposition (this behavior would persist no matter how many atoms $w_5, \ldots, w_n$ we would add to B, keeping the prior uniform over the whole B) and, regardless of the posterior $b'(R_1|R)$ imposed, leads to the decrease of the credence in R (see Appendix B of Seidenfeld (1986) for a proof). Here, as already said, we fix $b'(R_1|R)$ at $3/4$, but we are investigating the various options for $b'(B_1|B)$, that is, θ; it turns out that the more extreme (closer to 0 or 1) θ is, the higher the posterior

$b'(R)$ as mandated by MRE; for some θ, it will be higher than the original $b(R)$. The same is true for QUM.

Some might entertain the notion that, if MIRE gives the "correct" verdict by not requiring us to modify the credence in R, we may want to judge the relative qualities of MRE and QUM by how far they stray from what MIRE prescribes. Figure 5.2 makes it clear that this will not do: for some values of θ, QUM gives values of $b'(R)$ closer to 0.5 than the values given by MRE, but for some the situation is the opposite.

5.5.3. A nonuniform prior

Consider now a case in which the prior is non-uniform: $b(R_1) = 0.5$, $b(R_2) = 0.3$, $b(B_1) = 0.15$, $b(B_2) = 0.05$. Figure 5.3 portrays the behavior of the update methods we are dealing with in this case. The interpretation is as before.

Consider where the horizontal lines intersect the curves, that is, for what values of θ the update methods achieve the minimum inaccuracy in response to the a four-world version of the Judy Benjamin problem which resembles the original one in that it only has a single conditional, but departs from it in that it features our chosen non-uniform prior. MRE does this for $\theta = 3/4$; however, QUM achieves the minimum inaccuracy for θ close to 0.63. This means that for a non-uniform prior in response to the Judy Benjamin problem featuring only the posterior $b'(R_1|R)$, QUM would have us not only modify the credence in R, but also the conditional credence in B_1 given B, which MRE keeps intact.[16] This seems to be a slight advantage of MRE.

Note that it is clearly seen from Figure 5.3 that QUM, for some values of θ, would have us lower the credence in R below what it mandates for the original problem, with θ unspecified. This is a marked difference between QUM and MRE, which does not allow such situations.

[16] The already mentioned paper (Levinstein, 2012) reports similarly unwanted behavior of the AJC rule mandated by QUM.

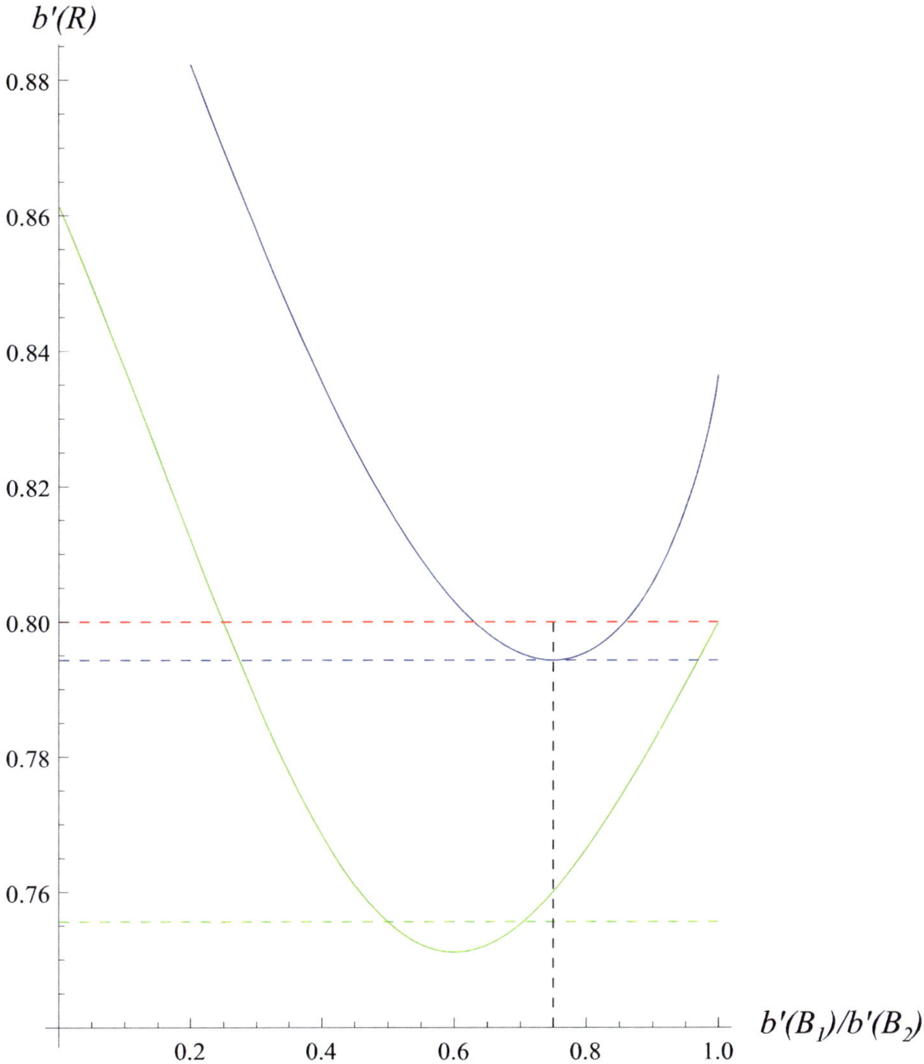

Figure 5.3. The symmetric JB problem for a nonuniform prior. (For the interpretation consult the text.)

5.6. Inverse Relative Entropy is a Bregman Divergence

I have been using the terms RE and IRE as they appear in Douven and Romeijn (2011); typically, whenever people speak about "relative entropy", they do not clearly specify the order of the arguments of the function they are concerned with, which leads to all sorts of confusions. Acknowledging this, I have followed this lead from Section 5.1.1 onwards; my task in this section is to describe and hopefully deal with some of them.

In a belief revision context, suppose your prior credence function is $\mathfrak{b}$ and you are looking for a posterior $\mathfrak{b}'$. If you would like to choose between the two formulas in the M(I)RE definitions on p. 101, the choice is effectively between having the posterior only inside the logarithm (in each summand), or both before and inside the logarithm (in each summand). The latter option has met with widespread approval; it is the one called "Infomin" by van Fraassen and "Minimising Relative Entropy" by Douven and Romeijn.[17] At this point I just wanted to rehearse the terminology I have adopted for this chapter, following Douven & Romeijn: if the posterior is only inside the logarithm, we are said to minimize the *inverse* relative entropy.

It would seem that notwithstanding Douven and Romeijn (2011), MIRE has not met with widespread approval. One of the reasons I have heard many times at conferences is that it is supposedly "not a Bregman divergence", that is, it does not belong to a class of functions that are the topic of a few practical theorems in Predd et al. (2009) and which have found philosophical application for example in the recent monograph Pettigrew (2016). This is not true. Actually, the *inverse* relative entropy function most definitely *is* a Bregman divergence, as we will shortly see. It is unclear, however, whether the same can be

[17] The labeling is a bit misleading as one might think that a method minimizing information should *maximize* entropy and vice versa. However, this will be the least of our problems, so I will leave this particular terminological issue aside. I will just mention here that Dziurosz-Serafinowicz (2015), on p. 1147, describes the method as that of "*maximum* relative entropy (MRE), also known as the rule of *minimizing* cross-entropy" (emphasis mine).

said about relative entropy! Later, after noting the connection between MRE, MIRE and the Kullback-Leibler divergences, I will point out that both MIRE and MRE are members of another class of divergences, the so-called f-divergences, which have found fruitful application in statistical inference, and so perhaps both should be welcome in formal epistemology.

To facilitate the later comparison with the literature on KL-divergences, we will modify our notation slightly.

Suppose $\mathbf{p} = \langle p_1, \ldots, p_n \rangle$ and $\mathbf{q} = \langle q_1, \ldots, q_n \rangle$ correspond to probabilistic mass functions, that is, they are vectors of probabilistic credences in members of an n-element partition; in other words, in each vector all the entries are nonnegative real numbers from the $[0, 1]$ segment which sum up to 1. Let us assume that $\mathbf{p}$ is the prior credence and that $\mathbf{q}$ is the posterior, that is, the variable. We can rewrite the two method definitions from p. 101 as follows:

Minimizing Relative Entropy (MRE), *rewritten*
Given evidential constraints C and the prior $\mathbf{p}$, the belief function $\mathbf{q}$ which should be adopted is such that it satisfies C and minimizes the expression

$$\mathsf{RE}(\mathbf{p}, \mathbf{q}) = \sum_{i=1}^{n} q_i \ln\left(\frac{q_i}{p_i}\right).$$

That is, the variable occurs *twice* in each summand, both before the logarithm and inside it.

Minimizing Inverse Relative Entropy (MIRE), *rewritten*
Given evidential constraints C and the prior $\mathbf{p}$, the belief function $\mathbf{q}$ which should be adopted is such that it satisfies C and minimizes the expression

$$\mathsf{IRE}(\mathbf{p}, \mathbf{q}) = \sum_{i=1}^{n} p_i \ln\left(\frac{p_i}{q_i}\right).$$

I will now argue that IRE is a Bregman divergence, in fact, an *additive* Bregman divergence.

Definition 12 (Additive Bregman divergence, Pettigrew (2016), p. 84): A function $D : [0,1]^n \times [0,1]^n \to [0,\infty]$ is an *additive Bregman divergence generated by* ϕ if:

- for all $\mathbf{x},\mathbf{y} \in [0,1]^n$ $D(\mathbf{x},\mathbf{y}) \geqslant 0$, with equality iff $\mathbf{p} = \mathbf{q}$ (divergence);
- there exists a function $d : [0,1]^2 \to [0,\infty]$ such that
$$D(\mathbf{x},\mathbf{y}) = \sum_{i=1}^{n} d(x_i, y_i);$$
- $\phi : [0,1] \to \mathbb{R}$ satisfies the following conditions:
  - ϕ is continuous, bounded, and strictly convex on $[0,1]$;
  - ϕ is continuously differentiable on $(0,1)$;
  - for all $x,y \in [0,1]$,
  $$d(x,y) = \phi(x) - \phi(y) - \phi'(y) \cdot (x-y)$$
  where the dot denotes multiplication and for $i \in \{0,1\}$ we define $\phi'(i) = \lim_{x \to i} \phi'(x)$.

Note that if D is an additive Bregman divergence generated by ϕ, we can put $\Phi(\mathbf{x}) = \sum_{i=1}^{n} \phi(x_i)$ and write the divergence as follows:

$$D(\mathbf{x},\mathbf{y}) = \Phi(\mathbf{x}) - \Phi(\mathbf{y}) - \nabla\Phi(\mathbf{y}) \cdot (\mathbf{x}-\mathbf{y})$$

where the dot is the dot product.

Let us check if we can construct IRE as a Bregman divergence. Choose $\phi(x) = x\ln(x)$ as the generating function.[18] Observe that $(x\ln(x))' = 1 + \ln(x)$. Notice that

$$d(x,y) = x\ln(x) - y\ln(y) - (1+\ln(y))(x-y) =$$
$$= x\ln(x) - y\ln(y) - (x\ln(y) + x - y\ln(y) - y) = y - x + x\ln\left(\frac{x}{y}\right).$$

[18] I do not wish to claim originality—this function is given in many sources, e.g., Basu et al. (2011), as the generating function of one of KL-divergences we will soon turn to.

Therefore, since the vectors are probabilistic and thus the entries sum to 1, we see that

$$D(\mathbf{x},\mathbf{y}) = \sum_{i=1}^{n} d(x_i, y_i) = \sum_{i=1}^{n} x_i \ln\left(\frac{x_i}{y_i}\right) = \mathrm{IRE}(\mathbf{x},\mathbf{y}).$$

Ergo, *inverse* relative entropy is a Bregman divergence with the generating function $x\ln(x)$. As for obtaining RE, the "popular choice", as a Bregman divergence, see the next subsection. Note that you can of course define, for a Bregman divergence $D(\mathbf{x},\mathbf{y})$, an *inverse* Bregman divergence $D^{I}(\mathbf{x},\mathbf{y}) := D(\mathbf{y},\mathbf{x})$. RE is of course an *inverse* Bregman divergence in this sense (with the generating function $x\ln(x)$).

5.6.1. Relation to Kullback-Leibler divergence and *f*-divergences

In formal epistemology it is customary to use the terms "relative entropy" and "Kullback-Leibler divergence" (KL-divergence for short) interchangeably; possibly because this has been the way information scientists have used the terms and how they are typically introduced in textbooks. However, usually not enough (if any) attention is paid to the fact that the order of arguments matters, even though it is common knowledge that the KL-divergence is not symmetric and does not satisfy the triangle inequality, so it is most decidedly not a metric. In a philosophical context, apart from computational results, we might really care about the functions which generate our inaccuracy measures (because they might for example be related to scoring functions with some known features). Therefore, it seems to me that it will pay off to minutely study the expressions we are dealing with on such occasions.

The following definition is from Cover and Thomas (2006), one of the popular textbooks in information theory (p. 19). The authors use logarithms with base 2; we will use the natural logarithm to continue the thread from the previous sections (this has no bearing on minimization, which is what we are concerned with).[19]

[19] Assume that we adopt the typical conventions that $0\ln(\frac{0}{0}) = 0\ln(\frac{0}{p}) = 0$ and $p\ln(\frac{p}{0}) = \infty$.

Definition 13 (Kullback-Leibler divergence): The Kullback-Leibler divergence of probabilistic mass vectors $\mathbf{p}$ and $\mathbf{q}$ is defined as follows:

$$\mathsf{KL}(\mathbf{p},\mathbf{q}) = \sum_{i=1}^{n} p_i \ln\left(\frac{p_i}{q_i}\right).$$

It seems to me that in the context of information theory and application of divergences to economics the above definition is widely adopted; that is, $\mathsf{KL}(\mathbf{p},\mathbf{q})$ denotes a function in which the p_i's come before the logarithm. The **reverse Kullback-Leibler divergence** of $\mathbf{p}$ and $\mathbf{q}$, which, if needed, we may write as $\mathsf{RKL}(\mathbf{p},\mathbf{q})$, is simply $\mathsf{KL}(\mathbf{q},\mathbf{p})$. If the arguments are not carefully chosen, these two expressions will have different values. It seems to me, then, that introducing a term D_{KL} as supposedly denoting the Kullback-Leibler divergence, but defining it so that $D_{KL}(\mathbf{p},\mathbf{q})$ equals $\mathsf{KL}(\mathbf{q},\mathbf{p})$ in the above (and common) sense, as is done for example by Gaifman and Vasudevan (2012, p. 150), hampers rather than promotes understanding and should be avoided.

Cover and Thomas, in a move which seems to be typical of the literature, introduce $\mathsf{KL}(\mathbf{p},\mathbf{q})$ as a divergence "between $\mathbf{p}$ and $\mathbf{q}$". Since KL-div is not symmetric, which seems also to be pointed out in every typical textbook, this of course has the consequence that the KL-div between $\mathbf{p}$ and $\mathbf{q}$ is different from the one between $\mathbf{q}$ and $\mathbf{p}$. A philosophical reader might wish to wonder, then, what the word "between" means in an information theory context, since it doesn't seem to be explained very often (if at all). Perhaps one could entertain asymmetric applications of "between" as in "The difference between 8 and 10 is 25%, while the difference between 10 and 8 is only 20%", but whether this is correct English, or introducing such ways of speaking to English might be fruitful, is not for me to judge. I am just puzzled about the insistence on using terms suggesting metricality as denoting something which most certainly is not a metric. Perhaps it is of some didactic help for information theory students, but my philosophical background precludes me from seeing it.

Other authors suggest using the terms "from" and "to", or "forward" and "backward". It is interesting that even extremely well-published and widely cited non-philosophical researchers differ in the application of these terms. For example, Csiszár (1991) considers $\mathsf{KL}(\mathbf{p},\mathbf{q})$ to be a divergence "of $\mathbf{p}$ from $\mathbf{q}$" (p. 2037–2038), while Abbas (2009) takes it as a divergence "from $\mathbf{p}$ to $\mathbf{q}$" (p. 26).

One thing which should be clear, if not completely trivial by now, and which I suspect at least some of the philosophers publishing on the topic might find a tiny bit surprising, is that we should realise that when minimising relative entropy $\mathsf{RE}(\mathbf{p}, \mathbf{q})$ (in the sense we have used in this work, which is the one in the well-known paper Douven and Romeijn (2011)) we are *not* minimizing the $\mathsf{KL}(\mathbf{p}, \mathbf{q})$, but rather the reverse Kullback-Leibler divergence of $\mathbf{p}$ and $\mathbf{q}$, that is, $\mathsf{KL}(\mathbf{q}, \mathbf{p})$.

We have already noted that it is easy to obtain $\mathsf{IRE}(\mathbf{p}, \mathbf{q})$, that is, $\mathsf{KL}(\mathbf{p}, \mathbf{q})$, as a Bregman divergence with the generating function $x \ln x$. More generally, for any real numbers m and k, the function $k + m \cdot x + x \cdot \ln x$ generates $\mathsf{KL}(\mathbf{p}, \mathbf{q}) = \mathsf{IRE}(\mathbf{p}, \mathbf{q})$ as a Bregman divergence. However, completing a similar task for the RKL seems to be a tricky matter. For example, Basu et al. (2011) write on p. 347 that with the generating function $x - \ln(x) - 1$ "a reverse Kullback-Leibler divergence is obtained". Note the "a"; if you carry out the calculation, you will see the reason for it. At this moment I do not know a generating function with would generate exactly $\mathsf{RKL}(\mathbf{p}, \mathbf{q})$ as a Bregman divergence. (I take this to suggest that philosophers using the term "Bregman divergence" in the context of entropy minimization problems should be doubly careful.)

However, both KL and RKL divergences belong without any reasonable doubt to a class of functions which has found wide usage in various areas of statistical inference (see the bibliography in Basseville (2013)): f-divergences.

Definition 14 (f-divergence (Basu et al. (2011), p. 342, Basseville (2013), p. 622): *f-divergences are all and only functions of the kind*

$$I_f(\mathbf{p}, \mathbf{q}) = \sum_{i=1}^{n} q_i f(\frac{p_i}{q_i}),$$

where $\mathbf{p}$ *and* $\mathbf{q}$ *are probability mass vectors and* $f(x)$ *is a convex function satisfying* $f(1) = 0$ *and which is strictly convex at* $x = 1$.[20]

[20] The two references given agree as to the form of the function, which will be important below, but I would like to mention here that the Basseville paper requires

It is easy to check that both the Kullback-Leibler divergence and the reverse Kullback-Leibler divergence are f-divergences, with the generating convex functions being $x \ln x$ and $-\ln x$, respectively.

Sadly, this field is also not free of confusion. Nielsen and Nock (2014) give (p. 3) the opposite assignment of the generating convex functions f to KL and RKL; that is, in their approach $-\ln x$ generates the KL divergence as an f-divergence and $x \ln x$ generates the RKL divergence. This is simply because they use the opposite order of arguments than Basseville when reporting on the definition of f-divergences; I am not judging, just noting the discrepancy (though Nielsen and Nock seem to be in a minority).

5.6.2. (Less intermediate) conclusions

It seems to me that it is not easy to form a philosophically satisfying, coherent picture out of this collection of variously chosen definitional conventions which lead to different roles played by different functions.

On the one hand, basic common sense seems to speak against defining f-divergences so that one and the same function $x \ln x$ generates the KL as a Bregman divergence, but generates the RKL as an f-divergence, as this, without some good additional arguments I was unable to locate, would seem to introduce confusion. This then speaks against the Nielsen and Nock option and for the more common way of defining things chosen for example by Basseville.

This, however, would have the consequence that the function $-\ln x$ —the only proper local scoring rule, which we know as the fundamental building block of the local-logarithmic inaccuracy measure from the previous chapter, and which, as we have seen, is intimately related to the method of minimizing $\mathsf{IRE}(\mathbf{p}, \mathbf{q})$, that is, $\mathsf{KL}(\mathbf{p}, \mathbf{q})$—would generate $\mathsf{RKL}(\mathbf{p}, \mathbf{q})$ as an f-divergence. This is a conceptual disparity which is hard to philosophically justify without some additional insight about the roles the generating functions play in the two types of divergences, insight I am unfortunately lacking.

---

of f some additional derivative-related features which are not met by some standard f-divergences, and which we will ignore here.

Neither of the two options seem entirely satisfactory, then, and the precise role played by the function $-\ln x$ still seems quite mysterious, at least to me. I hope that in the future, researchers more knowledgeable in information theory than me will step up to clear up these issues, and that what I have done in this chapter shows that there is genuine confusion regarding these matters in the literature, and it might be beneficial to philosophers in formal epistemology to double-check what exactly they are speaking about when they mention terms like "Kullback-Leibler divergence" or "relative entropy", since the results might surprise them.

I have also tried to collect arguments rehabilitating the method of minimizing inverse relative entropy. It seems to me that the fact it mandates Conditionalization and Jeffrey Conditionalization, its behavior with regard to the Judy Benjamin problem and its variants, how it faces the simultaneous update problem, the fact that it is a Bregman divergence, and possibly the fact that it uses an inaccuracy measure which is free of the so called "elimination counterexamples" (see Section 6.1 below) all speak in favor of it.

Appendix (proofs of facts from Section 5.3)

A note on notation: In all facts below we will be speaking about propositions A and B. There will be a finite number of possible worlds $w_1, \ldots, w_n$ (and in one case also w_0). Prior credences in these will be denoted by $v_1, \ldots, v_n$ (and in one case also v_0). For clarity (I hope) in the proofs below, we will denote posterior credences in singletons of elements of $A \cap \neg B$, in ascending order of the indices, by $a_0, \ldots, a_m$; in the trivial one-element case we will just use a. (In fact, below we will only explicitly write out one proof using just a_0 and a_1). The posterior credence in $B \cap \neg A$ will be denoted by b; that in $A \cap B$ by k (for *k*ommon), while that in $\neg A \cap \neg B$ by z. We will not use subscripts for b, k, and z, since in the proofs below the cases will be suppressed in which the three corresponding logical combinations of A and B are involved as consisting of more than one possible world; when written out in full, subscripts would be needed just as in the case of a.

Proof of Fact 8

Fact 8. *Let* $W = \{w_1, w_2, w_3, w_4\}$ *be the set of epistemically possible worlds for some agent. Let* A *be the proposition* $\{w_1, w_2\}$*; let* B *be* $\{w_2, w_3\}$*. Suppose that the agent's evidential constraints are that her posterior belief function* b' *satisfies* $b'(A) = p$ *and* $b'(B) = q$*, for some* p *and* q*. Let us label the number* $b(A \cap B) + \frac{p - b(A) + q - b(B)}{2}$ *as* K*. The belief update function which the agent should adopt according to QUM is fully determined by the two constraints and the following condition:*

$$b'(A \cap B) = \begin{cases} 0 & \textit{if } K < 0 \\ \min\{p, q\} & \textit{if } \min\{p, q\} < K \\ K & \textit{otherwise.} \end{cases}$$

Proof. The expected inaccuracy of a belief function b' as given by the quadratic measure is as follows:

$$\sum_{i=1}^{4} v_i \|w_i - b'\|^2$$

which reaches its minimum for the same argument as

$$\underbrace{(v_1 + \cdots + v_4)}_{1}(a^2 + \cdots + z^2) - 2v_1 a - 2v_2 k - 2v_3 b - 2v_4 z.$$

Using the equalities $a = p - k$, $b = q - k$, $z = 1 + k - p - q$ we can write this expression as a function of k only (since p and q are given). After expansion and some tedious but straightforward calculation (using also the equality $v_4 = 1 - (v_1 + v_2 + v_3)$), we see that it reaches its minimum for the same argument as

$$k^2 + (v_1 + v_3 - p - q)k. \tag{5.2}$$

The derivative of (5.2) is an increasing linear function reaching 0 at

$$k = \frac{p - v_1 + q - v_3}{2}$$

and so (5.2) is minimized for this value of k. We leave it to the reader to show that $k = K$ where K is as specified in Fact 8.

Now, the value of k as given by (5.2) might turn out to be negative or higher than p or q, and thus be unavailable to the agent. Since, as noted, the derivative of (5.2) is an increasing linear function, in these cases to minimize the expected inaccuracy the agent should set k to 0 or the lower of the values p and q, accordingly. □

Proof of Fact 9

Compared to the previous proof, this one adds one possible world to the proposition $A \cap \neg B$; we will index it as well as the prior credence in its singleton by 0.

Fact 9. *Let* $W = \{w_0, w_1, w_2, w_3, w_4\}$ *be the set of epistemically possible worlds for some agent. Let* A *be the proposition* $\{w_0, w_1, w_2\}$*; let* B *be* $\{w_2, w_3\}$*. Suppose that the agent's evidential constraints are that her posterior belief function* b' *satisfies the following:* $b'(A) = p = b(A)$ *and* $b'(B) = q$*, for some* p *and* q*. If the belief update function which the agent adopts is the one fully determined by the two constraints and the following condition*

$$\begin{aligned} b'(A \cap B) &= b(A \cap B) + 4/7(q - b(B)), \\ b'(\{w_0\}) &= b(\{w_0\}) - 2/7(q - b(B)), \\ b'(\{w_1\}) &= b(\{w_1\}) - 2/7(q - b(B)), \end{aligned}$$

then it is the one which QUM dictates the agent should adopt in this case.

Notice that the main claim of Fact 9 is (only) an implication: this is because here it is less convenient to deal with the cases in which it would seem from initial calculations that one's credence in a proposition should be negative than in the four-world case; we are not concerned with such examples. This is similar to the problems Leitgeb and Pettigrew have to deal with in their proof that QUM dictates that AJC is the method of updating in response to Jeffrey evidence. We are interested only in the "favorable" cases, because we want to use them to illustrate some deficiencies of the QUM method.

Notice also that this is a case of AJC updating: the change in credence in each element of $A \cap \neg B$ is obtained by decreasing the prior credence by the same number.

Proof. The change from the proof of Fact 8 is that instead of a single a, we now have to consider a_0 and a_1. The expected inaccuracy of the considered belief function reaches its minimum for the same argument as the following expression:

$$\underbrace{(v_0 + \cdots + v_4)}_{1}(a_0^2 + \cdots + z^2) - 2v_0a_0 - 2v_1a_1 - 2v_2k - 2v_3b - 2v_4z.$$

Using the equalities $a_0 = p - k - a_1$, $b = q - k$ and $z = 1 + k - p - q$ we can write this expression as a function of k and a_1 only. After expansion and some more tedious but still straightforward calculation, we see that it reaches its minimum for the same arguments as

$$2k^2 + (a_1 - 2p - 2q + 2v_0 + v_1 + 2v_3)k + a_1^2 + (v_0 - v_1 - p)a_1,$$

which, after using the equality $p = v_0 + v_1 + v_2$ (true since we assumed the the credence in A does not change), transforms into

$$2k^2 + (a_1 - 2q - v_1 - 2v_2 + 2v_3)k + a_1^2 + (-2v_1 - v_2)a_1. \tag{5.3}$$

(5.3) is a function of two variables, k and a_1. Its partial derivative w.r.t. a_1 reaches 0 for $a_1 = \frac{2v_1+v_2-k}{2}$; when we plug this into the partial derivative w.r.t. k and assume it equals 0 (since we are looking for a minimum) we arrive at

$$7k = 4q + 3v_2 - 4v_3$$

which is equivalent to

$$k = v_2 + 4/7(q - v_2 - v_3),$$

that is, to

$$b'(A \cap B) = b(A \cap B) + 4/7(q - b(B))$$

as required. The other two conditions of Fact 9, on $b'(\{w_0\})$ and $b'(\{w_1\})$, follow now by straightforward calculation (remember that the credence in A does not change). □

The next proof is a bit lengthy but does not use the method of Lagrange multipliers. Fact 10 contains a more approachable description of the belief update function which is fully specified in Fact 11. For that Fact, I give below a concise proof using Lagrange multipliers, so if the reader is familiar with the method, it might be better to go there (p. 141).

Proof of Fact 10

Fact 10. *Let* A *and* B *be two propositions such that* $A \cap B \neq \emptyset$ *and let* b *be the agent's prior belief function. Suppose that the agent's evidential constraints entail (only) her new credences* $b'(A)$ *and* $b'(B)$*. To arrive at the belief function* b' *that the agent should adopt according to MIRE, we should first calculate the agent's new credences in the logical combinations of* A *and* B *so that the following is true*

$$\frac{b(A \cap \neg B)}{b'(A \cap \neg B)} + \frac{b(\neg A \cap B)}{b'(\neg A \cap B)} = \frac{b(A \cap B)}{b'(A \cap B)} + \frac{b(\neg A \cap \neg B)}{b'(\neg A \cap \neg B)} \tag{5.4}$$

and then use Jeffrey Conditionalization.

Proof. First, in what follows we will use the following simple observation.

Observation 1. Suppose that for some natural number j, $x_0, \ldots, x_j$ and $y_0, \ldots, y_j$ are strictly positive real numbers. If $\forall i \in \{1, \ldots, j\} \frac{x_0}{y_0} = \frac{x_i}{y_i}$, then $\frac{x_0}{y_0} = \frac{\sum_{i=0}^{j} x_i}{\sum_{i=0}^{j} y_i}$.

Second, remember that it can be shown by a limit argument that according to MIRE the agent should use Jeffrey Conditionalization once the new credences in the four logical combinations of A and B are determined (see Pettigrew (2016) and Diaconis and Zabell (1982)).

We need to show that the relationship (5.4) holds independently of the number of atoms in the probability space. We will first show that it holds in the simplest four-world case, and then show that adding a finite number of worlds to each of the logical combinations of the two propositions does not break the relationship. Since the MIRE method is equivalent to the local-logarithmic one (p. 102), we will use the latter for calculations.

The four-world case. The local-logarithmic method tells us we need to minimize the following expression containing the single variable k:

$$-v_1 \cdot \ln(p-k) - v_2 \cdot \ln k - v_3 \cdot \ln(q-k) - v_4 \cdot \ln(1+k-p-q) \tag{5.5}$$

A modicum of analysis shows us the above expression reaches its minimum if k is chosen so that the following is true:

$$\frac{\nu_1}{a} + \frac{\nu_3}{b} = \frac{\nu_2}{k} + \frac{\nu_4}{1+k-p-q},$$

which is equivalent to (5.4).

One could be worried that such a k might be unavailable for the agent, that is, that it could be negative or higher than either p or q. Fortunately, this never happens. Assume without loss of generalization that $p \leqslant q$. The derivative of (5.5) is a continuous function of k in the segment $(0, p)$; with k approaching 0 it approaches negative infinity, while with k approaching p it approaches positive infinity, therefore by the intermediate value theorem it reaches 0 at a point inside the segment $(0, p)$. At that particular point, as analysis of the second derivatives shows, (5.5) reaches a minimum. The number k mandated by the logarithmic update method—and so by MIRE—is thus always available to the agent.

Cases with more worlds. We will consider what happens when we add more worlds to the proposition $A \cap \neg B$. In fact, the technique will, I hope, be evident after we deal with adding just a single world. To avoid much repetition, I will skip the three remaining cases of the other logical combinations of A and B.

As already mentioned, I believe it will be worthwhile to start with the case in which just a single world is added to $A \cap \neg B$; that is, assume that $W = \{w_0, w_1, w_2, w_3, w_4\}$, A is the proposition $\{w_0, w_1, w_2\}$, and B is $\{w_2, w_3\}$. Label $b'(\{v_0\})$ with a_0 and $b'(\{v_1\})$ with a_1; using the equality $a_0 = p - k - a_1$ we can write the expression to be minimized according to the local-logarithmic update method as

$$\begin{aligned} -\nu_0 \cdot \ln(p-k-a_1) - \nu_1 \cdot \ln a_1 - \nu_2 \cdot \ln k - \nu_3 \cdot \ln(q-k) + \\ -\nu_4 \cdot \ln(1+k-p-q). \end{aligned} \tag{5.6}$$

We will find expression (5.6)'s minimum when we calculate the values of a_1 and k for which its partial derivatives w.r.t. those variables equal 0 (and second derivatives are positive, checking which we leave to the reader). Now, (5.6)'s partial derivative w.r.t a_1 equals 0 iff

$$\frac{\nu_1}{a_1} = \frac{\nu_0}{p-k-a_1},$$

from which by Observation 1 we get that

$$\frac{\nu_0}{p-k-a_1} = \frac{\nu_0+\nu_1}{p-k}.$$

Now, expression (5.6)'s partial derivative w.r.t. k equals 0 precisely when

$$\frac{\nu_0}{p-k-a_1} + \frac{\nu_3}{q-k} = \frac{\nu_2}{k} + \frac{\nu_4}{1+k-p-q},$$

which we now see to be equivalent to

$$\frac{\nu_0+\nu_1}{p-k} + \frac{\nu_3}{q-k} = \frac{\nu_2}{k} + \frac{\nu_4}{1+k-p-q},$$

that is, to (5.4), as required.

Notice now that if we add not just one, but more worlds to $A\cap\neg B$, *nothing of essence changes*. Instead of just two partial derivatives, we need to consider more, but the information we receive is again that for each world in $A\cap\neg B$, the ratio of the old credence in its singleton to the new credence in its singleton is the same, and so is also the same (by Observation 1) as the ratio of $b(A\cap\neg B)$ to $b'(A\cap\neg B)$, at which point we turn to considering the partial derivative w.r.t. k and get (5.4) immediately.

Adding more worlds to the remaining three logical combinations of A and B changes, again, nothing of essence: we just need to consider more partial derivatives w.r.t. new variables, but what we learn is uniformly the information that inside the given logical combination of A and B the ratio of old credence to new credence is constant over singletons of all worlds, and (5.4) continues to hold regardless of how many (but finitely many) worlds we add. □

Proof of Fact 11

Fact 11. *Let A and B be two propositions such that $A\cap B\neq\emptyset$ and let b be the agent's prior belief function. Suppose that the agent's evidential constraints entail (only) her new credences $b'(A)=p$ and $b'(B)=q$. The belief function b' the agent should adopt according to MIRE is defined as follows:*

For any $i\in\{1,\dots,n\}$, $b'(\{w_i\}) = \frac{b(\{w_i\})}{-\lambda_0-\lambda_1\chi_{A,i}-\lambda_2\chi_{B,i}}$ with parameters λ_0, λ_1 and λ_2 determined by the following three equations:

$$\sum_{i=1}^{n} \frac{b(\{w_i\})}{-\lambda_0 - \lambda_1 \chi_{A,i} - \lambda_2 \chi_{B,i}} = 1;$$

$$\sum_{i=1}^{n} \chi_{A,i} \frac{b(\{w_i\})}{-\lambda_0 - \lambda_1 \chi_{A,i} - \lambda_2 \chi_{B,i}} = p;$$

$$\sum_{i=1}^{n} \chi_{B,i} \frac{b(\{w_i\})}{-\lambda_0 - \lambda_1 \chi_{A,i} - \lambda_2 \chi_{B,i}} = q.$$

Proof. The constraint $b'(A) = p$ is equivalent to $\sum_{i=1}^{n} \chi_{A,i} b'(\{w_i\}) = p$; $b'(B) = q$ is equivalent to $\sum_{i=1}^{n} \chi_{B,i} b'(\{w_i\}) = q$. Using the method of Lagrange multipliers (see, e.g., Cover and Thomas (2006)) we have:

$$\Lambda = \sum_{i=1}^{n} b(\{w_i\}) \ln\left(\frac{b(\{w_i\})}{b'(\{w_i\})}\right) - \lambda_0\left(\sum_{i=1}^{n} b'(\{w_i\}) - 1\right) -$$
$$\lambda_1\left(\sum_{i=1}^{n} \chi_{A,i} b'(\{w_i\}) - p\right) - \lambda_2\left(\sum_{i=1}^{n} \chi_{B,i} b'(\{w_i\}) - q\right).$$

Taking the partial derivative w.r.t. $b'(\{w_i\})$ we obtain

$$\frac{\partial \Lambda}{\partial b'(\{w_i\})} = -\frac{b(\{w_i\})}{b'(\{w_i\})} - \lambda_0 - \lambda_1 \chi_{A,i} - \lambda_2 \chi_{B,i}.$$

This equals 0 precisely when $b'(\{w_i\}) = \frac{b(\{w_i\})}{-\lambda_0 - \lambda_1 \chi_{A,i} - \lambda_2 \chi_{B,i}}$. Since the second derivatives are all positive, and our constraints define a closed and bounded set, we have arrived at a minimum. □

Chapter 6

Regarding the Brier Score

In Sections 5.2 and 5.3 of the previous chapter, I presented a few arguments which may be seen as suggesting that the Quadratic Update Method (QUM) is not best suited to some belief update problems. The QUM is a specific application of the so-called "Brier Score" (originally proposed by Brier (1950)), which is frequently used in formal epistemology in various forms. Since a major goal of the aforementioned section was to answer update problems posed in the specific framework of Leitgeb and Pettigrew (2010a and 2010b), I decided not to introduce the various forms the Brier Score might take in different applications. I believe the best introduction to that vast topic is the unpublished but publicly accessible work by Landes (2014).

While the main topic of the previous chapter was that of (minimizing (inverse)) relative entropy, and the argument against employing the QUM was something of a byproduct of this, the current concern is general arguments for and against using the Brier Score for assessing the value of an agent's cognitive state, in particular in the context of accuracy-first frameworks. Section 6.1 will concern the "elimination counterexamples" by Fallis and Lewis (2016) and Lewis and Fallis (2016), aimed at showing that one should not use the Brier Score to measure the epistemic utility of a credal state. I will argue that similar examples can be presented to argue against a few other inaccuracy measures, and so, that the Brier Score does not fair so badly as Lewis and Fallis would have it. In Section 6.2 I will study the arguments for using the QUM proposed by Leitgeb and Pettigrew (2010a): if QUM is not satisfactory, as I have argued, maybe we can spot some holes in these arguments? To the contrary, it turns out that the arguments by Leitgeb and Pettigrew can be strengthened, as I show in Section 6.2 below.

And so, the upshot of the current chapter is delicately opposite to the previous one: MIRE, while performing better in terms of belief updating, is *not* a clear winner in the context of accuracy-first frameworks.

6.1. Against: the elimination counterexamples

The QUM, which we encountered in the previous chapter, is closely related to a way of evaluating an agent's belief function called the "Brier Score"; effectively, it requires the agent to minimize the expected value of it. For the moment, following the previous chapter, assume that we are interested in the 'global' credence functions, that is, vectors of the agent's credences in 'singleton' propositions. This is a special case in which the set of propositions which is of interest to us forms a partition of the underlying set W; what we will say in this chapter will be independent of the choice of partition, and so we will make the choice of dealing with the most fine-grained one.

Assume that $W = \{w_1, \ldots, w_n\}$, w_i is the real world, and b, the belief function, is a probabilistic mass function over W, that is, recall, a vector where all the entries are nonnegative real numbers from the $[0, 1]$ segment which sum up to 1. In other words, b is a vector of credences in all members of a list of exclusive and jointly exhaustive hypotheses, interpreted as singleton sets of possible worlds. b_i is to be read as $b(\{w_i\})$. For such b's the following definition makes sense:

The Brier Score, partition version:

$$B(b) = (1 - b(\{w_i\}))^2 + \sum_{j=1, j \neq i}^{n} b(\{w_i\})^2.$$

That is, in this version the Brier Score takes into account the squares of differences between the given proposition's truth value (which equals 1 only in the case of $\{w_i\}$) and the agent's credence in it. Propositions different from singletons are entirely ignored.

Two other inaccuracy measures appropriate for partitions (Bickel (2007)):

- the logarithmic measure: $L(b) = -\ln(b(\{w_i\})$
- the spherical measure: $S(b) = -b(\{w_i\})/\sqrt{\sum_{j=1}^{n} b(\{w_j\})^2}$

However, we will also consider belief functions as defined over the whole set of propositions from $\mathcal{F} = \mathcal{P}(W)$, and not just the 'all the singletons' partition. In fact, the three measures about to be defined allow us to take an arbitrary set of propositions $\mathcal{F}$ and a belief function b defined on it. Like before, assume $\chi(A)$ is the truth value of the proposition A from $\mathcal{F}$.

Two inaccuracy measures appropriate (also) for whole Boolean algebras (Joyce (2009), p. 275):

- the additive logarithmic rule: $AL(b) = \sum_{A \in \mathcal{F}} -\ln(|(1-\chi(A)) - b(A)|)$
- the additive spherical rule: $AS(b) = \sum_{A \in \mathcal{F}} \frac{|(1-\chi(A))-b(A)|}{\sqrt{b(A)^2+(1-b(A))^2}}$

The Brier Score, a general version (Pettigrew (2016), (p. 36):

$$GB(b) = \sum_{A \in \mathcal{F}} (\chi(A) - b(A))^2$$

Fallis and Lewis (2016) argue that the Brier Score, at least in the partition version, is not a good tool for measuring the value of an agent's belief function.[1] The reason is that conditionalization is supposedly always of epistemic benefit to the agent, yet the Brier Score seems to disagree. That is, there are cases in which, according to the Brier Score, the inaccuracy of a belief function *increases* after conditionalization; we will label cases like this 'elimination counterexamples', slightly departing from Fallis and Lewis's terminology. This happens for example

[1] Fallis and Lewis actually use a slightly different version of the Brier score (partition version) than the one offered above, but the difference has no bearing on the argument, and the version used in the current text generalizes, I think, more intuitively to the context of Boolean algebras.

with the update from $\langle 0.25, 0.5, 0.25\rangle$ (Brier score $^7/_8$) to $\langle ^1/_3, ^2/_3, 0\rangle$ (Brier score $^8/_9$), with the first entry corresponding to the actual world. Note that the authors are concerned with the partition version of the score. It seems to me that a natural initial reaction to the counterexamples of Fallis and Lewis would be to say that excluding a false hypothesis leads to numerous changes in the agent's credences, especially if we consider those in other propositions instead of the ones belonging to the particular partition. If we take the whole Boolean algebra of the propositions the agent has an opinion about, then, after conditionalization, credences in some true propositions increase, credences in some true propositions decrease, and the same holds for false propositions, so the verdict should not be immediate. We will concisely consider Boolean algebras in Section 6.1.3 and will stick with partitions for now.

Note that the both the logarithmic and spherical measures are immune to elimination counterexamples. This is trivial for the logarithmic measure, since after conditionalization the credence in the true hypothesis increases, which is all that matters from the standpoint of this measure. That this holds also for the spherical rule is proven by Fallis and Lewis.

I would like to suggest a train of thought which, at least for me, dilutes the force of elimination counterexamples. In Chapter 5 we judged several update methods based on which update rules they lead to. However, one can reverse the tables, and ask for the justification of conditionalization (as opposed to other reactions to learning that some proposition is true; see for example the discussion of imaging immediately below). A natural thing to do in an inaccuracy-first framework is to point out that conditionalization minimizes *expected* inaccuracy (which is one way to describe some facts mentioned in 5.1.2). Notice that this does not entail (nor, to me, does it even *suggest*) that conditionalization will invariably decrease the inaccuracy—we presumably should have some other reason to expect this.

The epistemic benefit of conditionalization seems to come for the most part from excluding a false hypothesis. The credence formerly bestowed upon it needs to be transferred to the remaining hypotheses. This can of course be done in a number of ways. Conditionalization prescribes multiplying one's remaining credences by the same number so that their sum equals 1. This however seems to be unsuitable in many situations, some examples of which can be found for example in the first section of Perea (2009). David Lewis (1976) proposed a method he labelled 'imaging', which would have the agent transfer the whole credence from the excluded hypothesis to just one of the remaining ones, on the basis of (the agent's beliefs in) which possible world was most similar to the excluded one. This approach was generalized in Gardenfors (1982), and as far as the resulting 'general imaging' is concerned, there is no distribution of the credence in the excluded hypothesis among the remaining ones which would be *a priori* disallowed. For example, a version of imaging we will call 'uniform imaging' would let us *add* the same number to each credence in a remaining hypothesis so that their sum equals 1, which the reader might take to be quite close in spirit to the AJC rule discussed in the previous chapter.

Fallis and Lewis (Peter J.) are only concerned with conditionalization, but their argument might be strengthened if it can be shown that the Brier Score presents a similar behavior when faced with some reasonable alternative update rules. It is very easy to show that elimination counterexamples exist for imaging. For example, in the update from $\langle 0.25, 0.5, 0.25\rangle$ to $\langle 0.25, 0.75, 0\rangle$, which is an application of the original Lewisian imaging, the Brier score—if we assume that w_1 is the actual world—of the credence function increases from 7/8 to 9/8. However, we will prove that uniform imaging coupled with the Brier Score is immune to elimination counterexamples.

6.1.1. Uniform imaging

In this subsection we will continue to only deal with vectors where all the entries are nonnegative real numbers from the $[0, 1]$ segment which sum up to 1, that is, probabilistic mass functions: vectors of credences in all members of a list of exclusive and jointly exhaustive hypotheses, which, when needed, will be interpreted as singleton sets of possible worlds.

Assume that n hypotheses are entertained (at least 3), and the nth one is to be excluded, while the 1st one is true (assuming this leads to no loss of generality, since everything we will say holds for any choice of the true hypothesis).

Definition 15 (Update by uniform imaging): *A vector* $q = \langle q_1, \dots, q_n \rangle$ *is the* update by uniform imaging *of a vector* $p = \langle p_1, \dots, p_n \rangle$ $(p_n > 0)$ *if and only if:*

- $q_n = 0$;
- *for any* $i \in \{1, \dots, n-1\}$, $q_i = p_i + p_n/n-1$.

For example, $\langle 0.6, 0.4, 0 \rangle$ is the update by uniform imaging of $\langle 0.5, 0.3, 0.2 \rangle$, while conditionalization would lead from the same prior to $\langle 0.625, 0.375, 0 \rangle$. Of course, uniform imaging is equivalent to conditionalization if the prior is uniform.

We will now show that no case of update by uniform imaging can lead to elimination counterexamples, that is, the Brier Score of the posterior is always strictly lower than that of the prior.

Fact 12. *Suppose* q *is the update by uniform imaging of* p. *Then* $B(q) < B(p)$.

Proof. In the update from p to q, changes in two entries influence the Brier Score negatively: the last one (since the credence in the false hypothesis $\{w_n\}$ is transformed from p_n, which we assumed to be strictly positive, to 0) and the first one (since the credence in the true hypothesis $\{w_1\}$ is transformed from p_1 to $q_1 = p_1 + \frac{p_n}{n-1}$). So, the decrease in the Brier score is equal to $p_n^2 + (1-p_1)^2 - (1-(p_1 + \frac{p_n}{n-1}))^2 = p_n^2 + 2\frac{p_n}{n-1} - 2p_1\frac{p_n}{n-1} - \frac{p_1^2}{(n-1)^2} = (1 - \frac{1}{(n-1)^2})p_n^2 + 2(1-p_1)\frac{p_n}{n-1}$.

All the remaining $n-2$ entries offer an increase in the Brier score equal to $(p_i + \frac{p_n}{n-1})^2 - p_i^2 = \frac{p_n^2}{(n-1)^2} + 2p_i\frac{p_n}{n-1}$ in each case, and so the total increase in the Brier score equals $(n-2)\frac{p_n^2}{(n-1)^2} + \sum_{i=2}^{n-1} 2p_i\frac{p_n}{n-1}$. Since $p_n > 0$, we know that $\sum_{i=2}^{n-1} p_i < 1 - p_1$, and so $\sum_{i=2}^{n-1} 2p_i\frac{p_n}{n-1} < 2(1-p_1)\frac{p_n}{n-1}$. Since for natural $n > 2$ it holds that $n - 2 < n^2 - 2n$, we know that $(n-2)\frac{p_n^2}{(n-1)^2} < (1 - \frac{1}{(n-1)^2})p_n^2$. We see, then, that in the update from p to q the total increase in the Brier score is strictly lower than the total decrease, that is, that $B(q) < B(p)$. □

6.1.2. The case of the 12 drawers

Before we consider credence functions over Boolean algebras, I would like to show another update example in which not only *might* an exclusion of a false hypothesis increase the Brier Score, but in fact it is in some sense *expected* to happen: the agent, who obviously received beneficial evidence, would have to be exceptionally lucky to avoid the problem. Perhaps those who are worried by the elimination counterexamples will find that this strengthens their opinion on the subject. I do not want to put much stress on it, since, as the reader will see, various details of the story might be fleshed out so that it could support different arguments.

Consider the following situation: an agent is interested in the precise location of a document which she knows is in one of 12 drawers. The drawers are opened one at a time without the influence of the agent. The intelligence she has gathered leads her to credence 0.1 that the document is inside drawer number 1, credence 0.4 that it is inside drawer number 2, but leaves her indifferent with regard to the remaining drawers (which leads to 10 credences of 0.05 each). In reality the document is in drawer number 1. Suppose our agent uses conditionalization. Now, if *any* of the drawers 3 to 12 is opened, that is in 10 out of the 12 possible cases, the Brier score of the agent's credence *increases*, despite the epistemic benefit to the agent in the form of excluding a false hypothesis.

For example, if drawer number 12 is opened, then the initial credence $\langle .1, .4, .05, .05, .05, .05, .05, .05, .05, .05, .05, .05 \rangle$, whose Brier score is lower than 1, gets transformed via conditionalization into $\langle .105, .421, .053, .053, .053, .053, .053, .053, .053, .053, .053, 0 \rangle$, the Brier score of which is higher than 1.[2]

If the drawers are opened by a random device and each drawer has the same chance of being opened, our agent faces an 80% risk of lowering her Brier score despite conditionalizing after excluding a false hypothesis. Other similar examples can be easily constructed if relatively many false hypotheses are used.

[2] Yes, these fractions do not add up to 1—this is a result of rounding many entries up and few down.

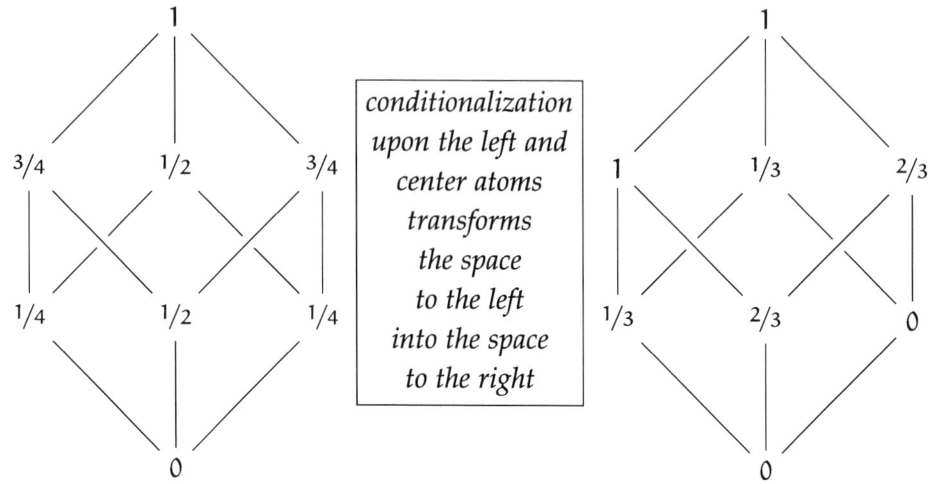

Figure 6.1. Conditionalization in the original "elimination counterexample" from Fallis & Lewis (2016): the "overall" inaccuracy also increases.

6.1.3. Credences over Boolean algebras

In this subsection I will show that in the context of Boolean algebras, conditionalization may lead to an increase of inaccuracy not only when it is measured by a version of the Brier score, but also if we opt for the additive logarithmic or additive spherical measures.

Consider, first, Figure 6.1, which depicts the original example from Fallis and Lewis, but in a version in which the belief function is defined over the whole Boolean algebra of propositions. The set $\mathcal{F}$ contains 8 propositions, two of which (the top and the bottom) are irrelevant since the credences in them are perfectly accurate. The true propositions are those in the upper left parallelogram. We can see that all changes in credences in true propositions have corresponding changes in the opposite direction (but with the same value) in credences in false propositions. Still, since quadratic functions are involved, the GB of the whole belief function increases from 1.75 to about 1.78. So, in a sense the message of the original elimination counterexample by Fallis and Lewis seems to be strengthened since it doesn't require us to restrict our attention to partitions: whole Boolean algebras also provide suitable examples.

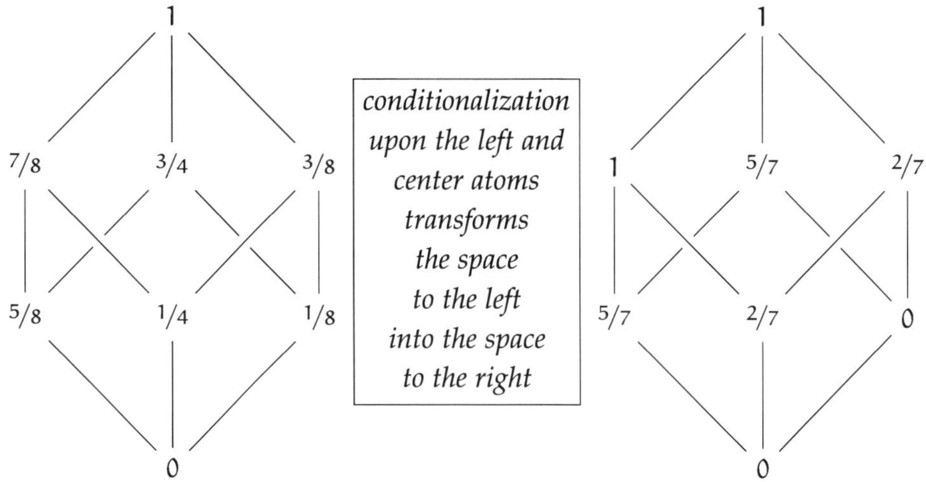

Figure 6.2. An example of conditionalization which features an increase in "overall" inaccuracy regardless of whether the inaccuracy measure of choice is the Brier score, the additive logarithmic measure, or the additive spherical measure.

However, the situation is not so simple. Figure 6.2 depicts a situation in which an agent increases her inaccuracy after conditionalization according to all three suitable measures we have mentioned: the General Brier score (from about 1.94 to about 2.04), the additive logarithmic measure (from about 5.00 to about 5.01), and the additive spherical measure (from about −3.64 to about −3.49). Examples like this abound; only rudimentary programming skills are needed to code a program which finds them via random search.

6.1.4. Elimination counterexamples: summary

If we restrict our attention to credences over partitions, it may seem that indeed there is something wrong with the Brier score as a measure of inaccuracy, or, more broadly, as a measure of the epistemic utility of one's credal state. Moreover, there are clear alternatives in the form of the spherical and logarithmic measures; this may be seen as another argument for an approach to belief update that requires the agent to minimize the inverse relative entropy, as advertised in the previous chapter, since, as was already mentioned, it is equivalent to the approach of minimizing the logarithmic measure.

However, when thinking about the merits and flaws of the Brier Score, there is no need to concern ourselves with partitions only. Conditionalization typically requires numerous changes in credences in propositions outside any given partition. As can be seen in examples like the one depicted in Figure 6.2, measures other than the Brier Score also lead to 'elimination counterexamples' in this context. One reaction to this could be that the Brier Score should not be singled out as the problematic measure since other measures seem to be equally problematic. Another reaction to which I am sympathetic would be, though, that this speaks against the idea of elimination counterexamples as providing insight about the value of inaccuracy measures. At best, they can serve as a reminder that there definitely has to be more to the value of one's credal state than a single accuracy-related number.

Consider for example a true proposition and its negation, and a change in the two corresponding credences from $\langle 0.25, 0.75 \rangle$ to $\langle 0.3, 0.7 \rangle$. In terms of the absolute value, the differences in credence are equal, yet the GB, AL and AS measures will score them differently. When evaluating the change in credences after conditionalization, we might care about different things: we might, e.g., prefer a measure which counts equally each change in a credence in some proposition which possesses the same numeric value (e.g., an increase of .05 should count equally regardless of whether the prior credence was .25 or .75) or, say, a measure which takes the same stance regarding the *ratios* (that is, an increase from 0.2 to 0.3 should count for as much as the one from 0.4 to 0.6). The situation seems to be similar to the one with confirmation measures (see, e.g., Eells & Fitelson (2002)); we might want to take into account how our inaccuracy measure treats different symmetries.

One could consider what might *prima facie* seem to be a route towards a better sort of argument against the Brier Score (partition version), or against any other inaccuracy measure in the more general version. Notice that all the examples discussed so far have dealt with the elimination of the last hypothesis $\{w_n\}$, assumed that the first one, $\{w_1\}$, was true, and observed that the measure in question led to an increase of inaccuracy after conditionalization. However, it is a matter of simple calculation that in all the displayed cases there was a different world w_m such that, were *that* world the actual one, the inaccuracy would decrease. While, when evaluating an inaccuracy measure, I am not moved too much by the fact that it allows that it *might* happen (or even, as in the case of the 12 drawers, that in some situations it is in some sense *likely* to happen) that the inaccuracy increases after conditionalization, I would take it as its real flaw if it allowed situations in which after conditionalization the inaccuracy increased *whatever the actual world was*. However, I was unable to find such examples for any of the inaccuracy measures discussed here, and I conjecture they do not exist. For the particular case of partitions, the case is already settled:

Fact 13. *There are no probabilistic mass vectors* p *and* q *such that* q *is obtained from* p *via conditionalization and* $B(p) < B(q)$, *regardless of the choice of the actual world* w_i.

(The conjecture was mine, and it has been recently shown to be true by Michał Tomasz Godziszewski; we are working on a joint paper on this and similar issues.)

In my opinion, Fact 13 acts as a sort of "sanity check" for the Brier Score: it is not *such* a bad measure of the value of one's credal state that it would permit conditionalization to be detrimental to that value, regardless of what the actual world is.

6.2. For: strengthening the arguments from the Ought-Can Principle

At this point the reader might be of the belief that the QUM is not the ideal belief update method, and—despite the arguments from the previous section—have some doubts whether the Brier Score is in general a good way of assessing the value of an agent's credal state.

Among the most extensively developed arguments for using quadratic inaccuracy measures are the three offered in Leitgeb and Pettigrew

(2010a).[3] If these lead to the QUM, and that method is inadequate, perhaps something is wrong with them? We will study this in this section. Examining the first two arguments will actually strengthen them (that is, we will arrive at the same conclusion from weaker assumptions), while after studying the third one I will propose an open question.

The arguments to be discussed are mathematical theorems interpreted in the context of the following two principles (Leitgeb and Pettigrew (2010a), p. 209):

Ought-Can Principle (OCP): A norm should not demand anything of an agent that is beyond her epistemic reach.

Accuracy: An epistemic agent ought to approximate the truth. In other words, she ought to minimize her inaccuracy.

The OCP goes back, arguably, to Kant (but see Stern (2004)); for a modern discussion see for example Vranas (2007). The norm of Accuracy is fleshed out in greater detail in Joyce (1998, p. 579).

The arguments given by Leitgeb and Pettigrew claim that if inaccuracy measures other than quadratic ones are used, in some situations the agent might end up facing a dilemma. That is, it is possible she will end up in a situation in which there are two reasonable ways of proceeding towards minimizing her inaccuracy which give different answers. Then, minimizing inaccuracy would be indeed "outside the epistemic reach" of the agent and the OCP would not hold. Assuming we are not satisfied with quadratic inaccuracy measures, at this point a reasonable conjecture would be, then, that the formal conditions which are to follow from the OCP are too stringent. Can we properly relax them so that more inaccuracy measures are admissible?

[3] Note also the recent book Pettigrew (2016), with a general argument for such inaccuracy measures appealing to some features of Bregman Divergences. I have to leave discussing this for the future.

6.2.1. The framework and the shared assumptions

For this chapter to be self-contained, and specifically so that the theorems make sense to the reader, I need to review the particulars of the LP framework in a more detailed manner than in Chapter 5. All definitions in this subsection are taken from Leitgeb and Pettigrew (2010a), and the page numbers given refer to that paper unless specified otherwise.

As before, assume we are given a finite set W of possible worlds. Each possible world w_i is identified with the ith unit vector in $\mathbb{R}^n$, that is, with a vector consisting of 0s and a single 1 in the ith place. *Belief functions* are mappings of the form $b : P(W) \to \mathbb{R}_0^+$. Bel($W$) is the set of all belief functions (note that no assumption of probabilism is made). *Global belief functions* are mappings of the form $b_{glo} : W \to \mathbb{R}_0^+$. Bel$_{glo}$($W$) is the set of all global belief functions.

Inaccuracy measures belong to one of two types, local or global, depending on whether they measure the inaccuracy of a credence in a specific proposition (given its truth value), or the inaccuracy of a whole belief function (from the perspective of a possible world), respectively. So, a *local inaccuracy measure* is a function $I : P(W) \times W \times \mathbb{R}_0^+ \to \mathbb{R}_0^+$. $I(A, w, x)$ is a measure of the distance of x from $\chi_A(w)$, the truth value of A at w. In contrast to this, a *global inaccuracy measure* is a function $G : W \times \mathrm{Bel}(W) \to \mathbb{R}_0^+$. $G(w, b)$ is a measure of the inaccuracy of a belief function b at world w.

Notice that since no assumption of probabilism is made (in fact, giving another justification for it is one of the aims of that paper), it is not obvious that the two types of inaccuracy measures are equally useful. Given a possible world, a global inaccuracy measure will give us the same verdict for all nonprobabilistic and the single probabilistic belief function compatible with a given b_{glo}, since global belief functions encode the agent's credences only in the singleton propositions. For this reason, one could claim that in cases in which local and global inaccuracy measures disagree, the local one should be given priority. It turns out that this is not the way to go: as proven at the very end of Leitgeb and Pettigrew (2010b), the norm of minimizing expected local inaccuracy cannot be satisfied in some quite natural situations (in which minimizing expected global inaccuracy is achievable). This fact somewhat offsets the just mentioned advantage that local inaccuracy measures seem to have over global ones. Therefore, in this section we will be treating local and global inaccuracy measures as philosophically on a par.

In view of the Accuracy norm, in response to incoming evidence which might influence the set E of worlds epistemically possible for the agent and entail constraints on belief functions, the agent with the current belief function b should adopt a credence in a proposition (in the local case) or a belief function (in the global case) which minimizes expected inaccuracy, defined as follows:

Definition 16 (Expected local inaccuracy, p. 206): *Given a local inaccuracy measure* I, *a belief function* b, *a degree of credence* x, *and propositions* $A, E \subseteq W$, *we define the* expected local inaccuracy *of degree of credence* x *in proposition* A *by the lights of* b, *with respect to* I, *and over the set* E *of epistemically possible worlds as follows:*

$$\mathrm{LExp}_b(I, A, E, x) = \sum_{w \in E} b(\{w\}) I(A, w, x).$$

Definition 17 (Expected global inaccuracy, p. 206): *Given a global inaccuracy measure* G, *belief functions* b *and* b', *and a proposition* $E \subseteq W$, *we define the* expected global inaccuracy *of* b' *by the lights of* b, *with respect to* G, *and over the set* E *of epistemically possible worlds as follows:*

$$\mathrm{GExp}_b(G, E, b') = \sum_{w \in E} b(\{w\}) G(w, b').$$

The key claim of Leitgeb and Pettigrew (2010a) is that the only "legitimate" inaccuracy measures—that is, the only measures which avoid certain problems when applied in the context of norms dictating that when updating her belief function the agent should minimize expected inaccuracy as defined above—are quadratic ones:

Legitimate inaccuracy measures (p. 218): The only **legitimate** inaccuracy measures are **quadratic** functions of $|\chi_A(w) - x|$ (in the local case) or $\|w - b_{glo}\|$ (in the global case).

As already mentioned, the first two arguments for this claim follow the same structure. A situation is proposed in which there are two methods of calculating the inaccuracy of a belief function, none of which is clearly preferable. These methods lead to numerically different results. Given that, the agent (supposedly) ends up in an epistemic dilemma, and thus the framework violates the OCP. A theorem is proven that a condition blocking this unwelcome consequence necessitates that quadratic inaccuracy measures are used.

We will discuss the two arguments in the next subsections. However, we should first state the four shared assumptions of these arguments (below, the expression "shared assumptions" refers to these four):

Local Normality and Dominance (p. 219): If I is a legitimate inaccuracy measure, then there is a **strictly increasing** function $f : \mathbb{R}_0^+ \to \mathbb{R}_0^+$ such that, for any $A \subseteq W$, $w \in W$, and $x \in \mathbb{R}_0^+$,

$$I(A, w, x) = f(|\chi_A(w) - x|).$$

Global Normality and Dominance (p. 219): If G is a legitimate global inaccuracy measure, then there is a **strictly increasing** function $g : \mathbb{R}_0^+ \to \mathbb{R}_0^+$ such that, for all worlds $w \in W$ and belief functions $b \in \text{Bel}(W)$,

$$G(w, b) = g(\|w - b_{glo}\|).$$

The Normality and Dominance conditions require simply that inaccuracy be nonnegative and increase with the distance from truth if the distance is understood as the absolute value of the difference, or the Euclidean norm. (I have to note here that Pettigrew currently believes that these assumptions already push us too far towards the intended conclusion regarding the quadratic inaccuracy measures—see the discussion of Theorem 3.2.2 in Pettigrew (2016).)

Local and Global Comparability (p. 220):

1. If $I(A, w, x) = f(|\chi_A(w) - x|)$ is a legitimate local inaccuracy measure, then $G(w, b) = f(\|w - b_{glo}\|)$ is a legitimate global inaccuracy measure.

2. If $G(w, b) = g(\|w - b_{glo}\|)$ is a legitimate global inaccuracy measure, then $I(A, w, x) = g(|\chi_A(w) - x|)$ is a legitimate local inaccuracy measure.

Assuming the previous two conditions, note that the functions f and g used in the Comparability condition are single argument functions from $\mathbb{R}_0^+$. That is, the functions are "oblivious" of the dimensions of the space containing the propositions involved, and so should either be fit for defining both local and global inaccuracy measures, or not fit for defining either.

Minimum Inaccuracy (p. 220):

1. If $I(A, w, x) = f(|\chi_A(w) - x|)$ is a legitimate local inaccuracy measure, then $f(0) = 0$.

2. If $G(w, b) = g(\|w - b_{glo}\|)$ is a legitimate global inaccuracy measure, then $g(0) = 0$.

The Minimum Inaccuracy condition simply conventionally sets the minimum of the inaccuracy function: in the best epistemic situation, when your credences match the real truth values, your inaccuracy should equal 0.

6.2.2. The argument from the "Discursive Dilemma"

The first argument (p. 222) concerns the situation in which, potentially, the sum of locally measured inaccuracies of an agent's credences regarding singleton propositions mismatches the globally measured inaccuracy of the whole measured function. That is, the inaccuracy of the global belief function b_{glo} does not equal the sum of local inaccuracies of $b(\{w_i\})$. Leitgeb and Pettigrew suggest that this would be analogous with some problems in judgment aggregation, and specifically the "Discursive Dilemma". The term is due to Pettit (2001), who notes that jurisprudential circles refer to it as the "doctrinal paradox".

One way of looking at the problem is the following. Assume you want to establish whether some propositions are true or false by taking a majority vote in a group consisting of an odd number of experts. Assume, also, that the experts adhere to classical logic. Suppose you are particularly interested in a certain compound proposition α. Then the information about the majority verdicts on all atomic propositions is not enough for you to infer the majority verdict on α. What's more, classical logic may view the majority verdicts on the atomic components of α as contradicting the majority verdict on α. Giving this a somewhat different spin, we can say that majority verdicts regarding premises do not in general entail majority verdicts about conclusions.

Pettit and Leitgeb and Pettigrew use a structurally identical example.[4] Three judges have to decide whether the accused is guilty or not. Assume that the accused is to be convicted iff propositions P and Q are true. Here are the judges' beliefs regarding P, Q, and whether the accused should be convicted, and also the verdict of the majority:

	P	Q	convict?
Judge 1	True	True	yes
Judge 2	True	False	no
Judge 3	False	True	no
Majority	True	True	yes / no

As we can see, "the majority", considered as an "aggregated" single agent, has trouble with classical logic, since the majority verdicts on P and Q are positive, while the majority verdict on the conjunction is negative. This, of course, in itself is not a problem—the issue becomes

4 Pettit continues the discussion with other cases, including for example disjunction.

troubling when we want to decide whether the convict should be sentenced.

(Not being an expert in jurisprudential disputes, I can offer an uninformed opinion I feel quite strongly about, namely, that—pragmatically—this is not such a big problem at all: the only thing to be done is to clearly specify the conditions for the conviction. If another agent is supposed to decide on the basis of the three opinions of the Judges, then that agent may care only about the Judges' majority verdicts regarding P and Q, and construe the verdict about $P \wedge Q$ herself. If, however, we wish to say that the conviction is to be decided solely on the basis of the majority verdict regarding $P \wedge Q$, then we may not care about the majority verdicts regarding the constituent propositions. Still, I agree that—conviction or no conviction—we might face a lingering sense of moral doubt.)

That example serving as a backdrop, Leitgeb and Pettigrew claim that if measuring the global inaccuracy of a belief function yields a different result than summing up the local inaccuracies of credences that belief function offers regarding the singleton propositions, then the agent in question faces an irresolvable dilemma, and so, if the framework allows such cases, then it violates the Ought-Can principle. Before proceeding with the formal analysis of a strengthening of Leitgeb and Pettigrew's argument, I have to note two issues I have regarding that idea.

First, it is not clear to me that the aforementioned agent will have a dilemma at all. True, she will arrive at different conclusions if she attempts to calculate what would appear to be the same thing—that is, the inaccuracy of a set of credences regarding atomic propositions—in two ways. Still, not only is it not obvious that, epistemically, a global belief function and a set of credences in singleton propositions are "the same thing", but also the different values at which the agent arrives may have no bearing regarding belief update problems. That is, the two procedures might recommend the same way of minimizing one's accuracy.

Second, there is a bijection between the set of all global belief functions on some finite W and the set of sets of credences in all singleton propositions from $\mathcal{P}(W)$: it is obvious how to go from one to the other and back. In contrast to that, a "majority belief function" contains *less* information than the set of the belief functions of the judges. If there is some better way of describing the analogy, unfortunately I do not see it forthcoming.

Putting these doubts aside, let us suppose that the analogy holds and so the agent in the aforementioned situation may really face a dilemma if the set of legitimate inaccuracy measures is not sufficiently restricted. Leitgeb and Pettigrew show that any possibility for the occurrence of such a dilemma is blocked by the following additional assumption:

Agreement on Inaccuracy (p. 223): Suppose I is a legitimate local inaccuracy measure. Then, by Local Normality and Dominance, there is a strictly increasing function $f : \mathbb{R}_0^+ \to \mathbb{R}_0^+$ such that $I(A, w, x) = f(|\chi_A(w) - x|)$. Further, by Local and Global Comparability, $G(w, b) = f(\|w - b_{glo}\|)$ is a legitimate global inaccuracy measure. Therefore, the following must hold: if b is a belief function and w_i is a world,

$$G(w_i, b) = \sum_{j=1}^{n} I(\{w_j\}, w_i, b(\{w_j\})).$$

That is,

$$f(\|w_i - b_{glo}\|) = \sum_{j=1}^{n} f(|\chi_{\{w_j\}}(w_i) - b(\{w_j\})|).$$

Leitgeb and Pettigrew then show that with the assumption in place, the only legitimate inaccuracy measures are quadratic functions:

Theorem (Theorem 3 from Leitgeb & Pettigrew (2010a)). *The following propositions are equivalent:*

1. *Function f is strictly increasing, and for all belief functions b and worlds w_i,*

$$f(\|w_i - b_{glo}\|) = \sum_{j=1}^{n} f(|\chi_{\{w_j\}}(w_i) - b(\{w_j\})|).$$

2. *There is a $\lambda \in \mathbb{R}_{>0}$ such that, for all $x \in \mathbb{R}_0^+$, $f(x) = \lambda x^2$.*

However, is imposing Agreement on Inaccuracy the way to go to avoid the Discourse-like dilemmas?

Remember that the norm of Accuracy says that the agent should *minimize* the (expected) inaccuracy. The agent will not really face any such dilemma if the two ways of calculating inaccuracy impose *the same ordering* on the considered belief functions, thus offering the same minima; that their verdicts should be numerically identical is *unnecessary*. Maybe it is precisely this addition which allows Leitgeb & Pettigrew to derive their intended conclusion?

Let us consider the following apparent weakening of the condition, which will require only that the two ways of calculating inaccuracy *agree on their verdict* when asked which of some two belief functions is more inaccurate:

Agreement* on Inaccuracy: Suppose I is a legitimate local inaccuracy measure. Then, by Local Normality and Dominance, there is a strictly increasing function $f : \mathbb{R}_0^+ \to \mathbb{R}_0^+$ such that $I(A, w, x) = f(|\chi_A(w) - x|)$. Further, by Local and Global Comparability, $G(w, b) = f(\|w - b_{glo}\|)$ is a legitimate global inaccuracy measure. Therefore, the following must hold: if b^1 and b^2 are belief functions and w_i is a world,

$$f(\|w_i - b^1_{glo}\|) < f(\|w_i - b^2_{glo}\|)$$

iff

$$\sum_{j=1}^{n} f(|\chi_{\{w_j\}}(w_i) - b^1(\{w_j\})|) < \sum_{j=1}^{n} f(|\chi_{\{w_j\}}(w_i) - b^2(\{w_j\})|).$$

Agreement* on Inaccuracy ensures that both ways of calculating inaccuracy will arrive at the same minima, or at the same minimum—if it is unique—and so the OCP would not be violated. In other words, Agreement on Inaccuracy requires that the two ways of calculating inaccuracy *exactly coincide* in values, which, according to Theorem 3 from Leitgeb and Pettigrew (2010a), leaves only the quadratic functions as legitimate inaccuracy measures. In contrast, Agreement* on Inaccuracy requires only that the two ways impose *the same ordering* on the considered belief functions. A legitimate question is, then, whether using the apparently weaker condition leaves us with more than just quadratic functions.

It turns out that the answer to this question is negative, due to the following theorem:

Theorem 5. *Suppose that the function* f *satisfies Local and Global Normality and Dominance, and Local and Global Comparability. Then the following propositions are equivalent:*

1. *Function* f *is strictly increasing and satisfies Agreement* on Inaccuracy;*
2. *there is a* $\lambda \in \mathbb{R}_{>0}$ *such that for all* $x \in \mathbb{R}_0^+$, $f(x) = \lambda x^2$.

Proof. To show our Theorem 5 we will use a modified version of the argument from Leitgeb and Pettigrew (2010a, p. 231). We will argue that the following conditions are equivalent:

i) Function g is strictly increasing and, for all belief functions b and worlds $w_i \in W$,

$$g(\|w_i - b^1_{glo}\|^2) \leqslant g(\|w_i - b^2_{glo}\|^2)$$

iff

$$\sum_{j=1}^{n} g(|\chi_{\{w_j\}}(w_i) - b^1(\{w_j\})|^2) \leqslant \sum_{j=1}^{n} g(|\chi_{\{w_j\}}(w_i) - b^2(\{w_j\})|^2);$$

ii) There is a $\lambda \in \mathbb{R}_{>0}$ such that for all $x \in \mathbb{R}_0^+$, $g(x) = \lambda x$.

That ii) implies i) follows straight from the third paragraph of the proof Leitgeb and Pettigrew give on p. 231 of their paper. Since, as they show, an increasing function satisfies Agreement on Inaccuracy, from which Agreement* on Inaccuracy follows, then an increasing function has to satisfy Agreement* on Inaccuracy also. (For a $\lambda \in \mathbb{R}_{>0}$ the function $g(x) = \lambda x$ is of course increasing.)

We now turn to showing that i) implies ii). For a *reductio*, suppose, then, that g satisfies i) but not ii), and so that g is increasing but not linear. It follows that there exist four real numbers a, b, c, and d such that $b - a = d - c$ but $g(b) - g(a) \neq g(d) - g(c)$. Notice that $g(b + c) = g(a + d)$, since g is a function. We will show that $g(b + c) \neq g(a + d)$, thus obtaining a contradiction.

Without loss of generality let us assume that $g(b) - g(a) > g(d) - g(c)$ (the proof for the other direction is analogous). Set W as $\{w_1, w_2\}$. Define the two belief functions b^1 and b^2 on the singletons as follows: $b^1(\{w_1\}) = 1 - \sqrt{a}$, $b^1(\{w_2\}) = \sqrt{d}$, $b^2(\{w_1\}) = 1 - \sqrt{b}$, $b^2(\{w_2\}) = 1 - \sqrt{c}$. Therefore the functions b^1_{glo} and b^2_{glo} are defined; we see that $a + d = \|w_1 - b^1_{\text{glo}}\|^2$ and $b + c = \|w_1 - b^2_{\text{glo}}\|^2$. Noting that:

$$a = |\underbrace{\chi_{\{w_1\}}(w_1)}_{1} - b^1(\{w_1\})|^2,$$

$$d = |\underbrace{\chi_{\{w_2\}}(w_1)}_{0} - b^1(\{w_2\})|^2,$$

$$b = |\underbrace{\chi_{\{w_1\}}(w_2)}_{0} - b^2(\{w_1\})|^2,$$

$$c = |\underbrace{\chi_{\{w_2\}}(w_2)}_{1} - b^2(\{w_2\})|^2$$

we see that from $g(b) - g(a) > g(d) - g(c)$, that is, from $g(a) + g(d) < g(b) + g(c)$, using i) we arrive at $g(a + d) < g(b + c)$, contradicting our previous assertion that $g(b + c) = g(a + d)$. □

Note that, in contrast to the proof Leitgeb and Pettigrew give for their result, we did not need to appeal to Cauchy's theorem about the only additive monotone functions on the reals being linear.

To sum up, in this subsection we have actually *strengthened* the argument from the "Discursive Dilemma" given in Leitgeb and Pettigrew (2010a) for limiting the set of legitimate inaccuracy measures to only the quadratic ones. Still, if we are not convinced the dilemma is a real one, or that it translates properly to the geometric inaccuracy-based epistemological framework we are dealing with, we might not be persuaded. Leitgeb and Pettigrew offer two more arguments which will be discussed in the two following subsections.

6.2.3. The argument from separability of global inaccuracy

The second argument uses the notion of a projection of a vector $(a_1, \ldots, a_n)$ on an axis j ($proj_j((a_1, \ldots, a_n))$) or a set of axes D ($proj_D((a_1, \ldots, a_n))$). Suppose $1 \leqslant j \leqslant n$; $D \subseteq \{1, \ldots, n\}$; $(a_1, \ldots, a_n) \in \mathbb{R}_0^n$. Then:

$$proj_j((a_1, \ldots, a_n)) = (a_1, \ldots, a_{j-1}, 0, a_{j+1}, \ldots, a_n)$$

$$proj_D((a_1, \ldots, a_n)) = (a_k^{'})_{k \in \{1, \ldots, n\}} : a_k^{'} = \begin{cases} 0 & \text{if } k \in D \\ a_k & \text{if } k \notin D. \end{cases}$$

Suppose a world w_i is given. According to Leitgeb and Pettigrew (p. 224), there seem to be[5] two ways of measuring the inaccuracy of the agent's global belief function $(a_1, \ldots, a_n)$, which use G or a combination of G and I:

- simply $G(w_i, (a_1, \ldots, a_n))$;
- for any w_j with $i \neq j$, take

$$I(\{w_j\}, w_i, a_j) + G(proj_j(w_i), proj_j((a_1, \ldots, a_n))).$$

In the second case we consider the subspace spanned by $W - \{w_j\}$ and divide the initial global belief function into two parts: a projection on that subspace, the inaccuracy of which we calculate using a global inaccuracy measure, and the credence in the singleton $\{w_j\}$, for calculating the inaccuracy of which we use a local inaccuracy measure. We then sum the two numbers.

If these two ways of calculating inaccuracy give different results, then, supposedly, a dilemma might arise for the agent. Leitgeb and Pettigrew propose blocking this possibility using the following condition:

Separability of Global Inaccuracy (p. 225): Suppose I is a legitimate local inaccuracy measure. Then, by Local Normality and Dominance, there is a strictly increasing function $f : \mathbb{R}_0^+ \to \mathbb{R}_0^+$ such that $I(A, w, x) = f(|\chi_A(w) - x|)$. Further, by Local and Global Comparability, $G(w, b) = f(||w - b_{glo}||)$ is a legitimate global inac-

[5] Regarding the "seem" see below.

curacy measure. Then, the following must hold: for all $w_i, w_j \in W$ with $i \neq j$,

$$G(w_i, (a_1, \ldots, a_n)) = f(\|w_i - (a_1, \ldots, a_n)\|) =$$

$$= f(|\chi_{w_j}(w_i) - a_j|) + f(\|proj_j(w_i) - proj_j((a_1, \ldots, a_n))\|).$$

It is again shown (Theorem 4, p. 231 of Leitgeb and Pettigrew (2010a)) that together with the shared assumptions presented above, this entails the only legitimate inaccuracy measures being quadratic functions.

But it is not evident what kind of a dilemma is made possible by the difference in value of the two ways of calculating inaccuracy.

First, we might agree that there are in general two ways of measuring inaccuracy: the global one and the local one. It is not evident that any combination of the two, and in particular, the combination the authors propose, should be equally good. I believe some argument *pro* should be given.

Second, it is again *a priori* possible that the ways do not exactly coincide in values, but give the same answers to all questions of the sort "is function b_1 better than b_2?" or, in general, "which belief function(s) from some given set minimize (expected) inaccuracy?". And so, this apparent weakening of the Separability Condition might seem reasonable:

Separability* of Global Inaccuracy: Suppose I is a legitimate local inaccuracy measure. Then, by Local Normality and Dominance, there is a strictly increasing function $f : \mathbb{R}_0^+ \rightarrow \mathbb{R}_0^+$ such that $I(A, w, x) = f(|\chi_A(w) - x|)$. Further, by Local and Global Comparability, $G(w, b) = f(\|w - b_{glo}\|)$ is a legitimate global inaccuracy measure. Then, the following must hold: for all $w_i \in W$ and b^1, $b^2 \in Bel(W)$:

$$f(\|w_i - b^1_{glo}\|) < f(\|w_i - b^2_{glo}\|)$$

iff

for any $D, E \subset \{1, \ldots, n\}$, $i \notin D \cup E$:

$$\sum_{j\in D} f(|\chi_{w_j}(w_i) - b^1(\{w_j\})|) + f(\|proj_D(w_i) - proj_D(b^1_{glo})\|) <$$

$$< \sum_{j\in E} f(|\chi_{w_j}(w_i) - b^2(\{w_j\})|) + f(\|proj_E(w_i) - proj_E(b^2_{glo})\|).$$

The original Separability condition deals with a single axis. But since it requires the equality of the inaccuracy values as calculated in the two ways, it generalizes to any finite set of axes. In the case of Separability*, though, since we only require the inequality, we need to be more general from the start.

Due to the following theorem, the apparent weakening is again, however, only really apparent:

Theorem 6. *Suppose that the function f satisfies Local and Global Normality and Dominance, Local and Global Comparability, Minimum Inaccuracy. Then the following propositions are equivalent:*

i) *Function f satisfies Separability* of Global Inaccuracy;*

ii) *there is a $\lambda \in \mathbb{R}_{>0}$ such that, for all $x \in \mathbb{R}^+_0$, $f(x) = \lambda x^2$.*

Proof. Since Separability* of Global Inaccuracy is weaker than Separability of Global Inaccuracy, and we know from Theorem 4 from Leitgeb and Pettigrew (2010a) that quadratic functions satisfy the latter, then we know that quadratic functions satisfy the former.

As for whether i) implies ii), assume that f satisfies Separability* of Global Inaccuracy. Choose an $i \in \{1, \ldots, n\}$ and define $D = E = \{1, \ldots, n\} \setminus \{i\}$. Notice that f satisfies Agreement* on Inaccuracy, and so, by Theorem 5, is a quadratic function. □

It turns out, then, that also in this case we have strengthened one of Leitgeb and Pettigrew's arguments for exclusively using the quadratic functions as legitimate inaccuracy measures.

6.2.4. The argument from directed urgency

Leitgeb and Pettigrew propose (p. 226) that the agent should be able to assess:

- the urgency of abandoning a given credence by the lights of some belief function;
- the direction in which that credence is to be changed;
- the relative urgencies of abandoning different credences / belief functions.

These would be provided by the derivative of, say, $\mathrm{LExp}_b(I, A, E, x)$ with respect to x. We need to assume such derivatives exist:

> **Continuous Differentiability (p. 226):**
>
> 1. If $I(A, w, x) = f(|\chi_A(w) - x|)$ is a legitimate local inaccuracy measure, then f is continuously differentiable on $\mathbb{R}_0^+$.
> 2. If $G(w, b) = g(\|w - b_{glo}\|)$ is a legitimate global inaccuracy measure, then g is continuously differentiable on $\mathbb{R}_0^+$.

According to Leitgeb and Pettigrew, the values of these derivatives should again coincide:

> **Agreement on Directed Urgency (p. 228):** If $I(A, w, x) = f(|\chi_A(w) - x|)$ and $G(w, b) = f(\|w - b_{glo}\|)$ are legitimate local and global inaccuracy measures, respectively, and f is differentiable, then, for all belief functions b and b' and all worlds $w_j \in W$,
>
> $$\frac{d}{dx}\mathrm{LExp}_b(I, \{w_j\}, E, x) =$$
>
> $$= \frac{d}{dx}\mathrm{GExp}_b(G, E, (b'(\{w_1\}), \ldots, b'(\{w_{j-1}\}), x, b'(\{w_{j+1}\}), \ldots, b'(\{w_n\})).$$

As the reader might anticipate at this point, Leitgeb and Pettigrew show (Theorem 5, p. 232) that adopting this entails that the legitimate inaccuracy measures are quadratic functions. However, we should again ask: why should the agent be interested in *actual values* of the derivatives? Does the value "7" mean the change needs to be urgent, or not?

The epistemic worth of the derivatives (apart from their signs) seems to lie in their *relative* value. But how exactly should we interpret $\frac{d}{dx}\mathrm{LExp}_b(I, \{w_j\}, E, x)$, if $x \neq b(\{w_j\})$?

- If b is our current belief function: this is our urgency of changing our credence in $\{w_j\}$ *had it been different* (or what we believe to be the urgency with which somebody else should change their credence in $\{w_j\}$);
- if b is not our current belief function, but x is our current credence in $\{w_j\}$: this is our urgency of changing our credence in $\{w_j\}$ had our belief function been not as it actually is.

The point that what can be useful are the *relative* values of the derivatives already suggests a similar transformation of the condition proposed by Leitgeb and Pettigrew as in the previous cases. However, I would like to consider a related but different question. Assume we are to adopt a new belief function b but we cannot do it instantaneously and need to proceed proposition by proposition. Which ones should we start with? A similar case would be a situation in which we want to give advice to another person regarding in which propositions she should first update her credence.

It seems reasonable that for two propositions $\{w_j\}$ (with credence r) and $\{w_k\}$ (with credence s), we should, then, compare (the absolute values of)

$$\frac{d}{dx}\mathrm{LExp}_b(I, \{w_j\}, E, x)\Big|_{x=r}$$

with

$$\frac{d}{dx}\mathrm{LExp}_b(I, \{w_k\}, E, x)\Big|_{x=s}$$

and the verdict should be the same had we used GExp_b. This motivates the following condition:

Agreement on Directed Urgency**: If $I(A, w, x) = f(|\chi_A(w) - x|)$ and $G(w, b) = f(\|w - b_{glo}\|)$ are legitimate local and global inaccuracy measures, respectively, and f is differentiable, then, for all belief functions b and b', worlds $w_j, w_k \in W$, and $r, s \in \mathbb{R}_0^+$,

$$\left|\frac{d}{dx}\text{LExp}_b(I, \{w_j\}, E, x)\Big|_{x=r}\right| \leqslant \left|\frac{d}{dx}\text{LExp}_b(I, \{w_k\}, E, x)\Big|_{x=s}\right|$$

iff

$$\left|\frac{d}{dx}\text{GExp}_b(G, E, (b'(\{w_1\}), \ldots, b'(\{w_{j-1}\}), x, b'(\{w_{j+1}\}), \ldots, b'(\{w_n\}))\Big|_{x=r}\right|$$

$$\leqslant$$

$$\left|\frac{d}{dx}\text{GExp}_b(G, E, (b'(\{w_1\}), \ldots, b'(\{w_{k-1}\}), x, b'(\{w_{k+1}\}), \ldots, b'(\{w_n\}))\Big|_{x=s}\right|$$

One could certainly consider variants of this condition, e.g., ones not referring to absolute values.

While I believe that my condition is better motivated than the one proposed by Leitgeb and Pettigrew, I was not able to obtain any hard results regarding the issue of the set of legitimate inaccuracy measures it delineates. Therefore, I would like to propose the following open problem:

Problem 1. *Which functions satisfy the shared assumptions, Continuous Differentiability and Agreement** on Directed Urgency?*

*

In this section I have discussed the 3 arguments Leitgeb and Pettigrew give for using exclusively quadratic functions as inaccuracy measures. I believe I have strengthened the first two arguments, but the authors themselves think the last one is the best (p. 226). I suggest that an answer to Problem 1, according to which measures other than quadratic ones would also be admitted as legitimate, could, then, trump the previous arguments. The answer to that problem, though, remains to be found.

Conclusion: formal justificational pluralism

It is *prima facie* reasonable to think that the same method of evaluating the value of one's belief function, e.g., measuring its inaccuracy, should be used when arguing for both synchronic and diachronic norms of rationality. For example, the same measure of accuracy could be appealed to when arguing for Probabilism, which is a synchronic norm, and, in the context of a minimization argument, used when backing up Conditionalization, which is a diachronic one.

However, the last two chapters suggest a different route. While I believe some of the recent arguments *against* the Brier Score—the ones using so-called "elimination counterexamples"—do not hold water, I think I have also strengthened some arguments *for* using it as a measure of inaccuracy. This seems to suggest that the Brier score is a legitimate tool for arguing for synchronic norms of rationality.

Chapter 5, though, pulls in the opposite direction. I tried to show there that applying (a form of) the Brier Score to several update problems leads to deficient conclusions and that using the Inverse Relative Entropy fares better (also in comparison with employing Relative Entropy) with regard to all the problems considered. This suggests that until a belief update problem is proposed to which MIRE gives a clearly bad answer, it is reasonable to consider it a viable tool for belief update problems. However, in light of the results from Section 6.2, I do not wish to hold that the local logarithmic scoring rule (the one minimizing which is equivalent to using MIRE) is the way to measure the inaccuracy of one's credence function in the context of arguments for synchronic norms. It would seem to follow that I should not claim that, if an inaccuracy measure is settled upon, then answers to belief update problems should appeal to minimizing the expected inaccuracy thusly measured. One simple reason for this may be that since a belief update problem asks for revision of your prior, then calculating the expected inaccuracy from the perspective of that prior—the one you know you need to abandon in order not to run afoul of the evidence you have gained—does not seem too reasonable. However, I do not see a clear alternative which would be generally preferable.

The failure of the measure supported by arguments appealing to the intended features of inaccuracy measures from the last section in application to belief update problems (Chapter 5) suggest the following position with regard to epistemic norms of credence, which I propose be labelled as *formal justificational pluralism*: it is possible that methods which may be used when arguing for synchronic norms are different from those which may be used when arguing for diachronic norms. I suggest using the Brier Score for synchronic norms and Minimizing Inverse Relative Entropy for diachronic ones.

Bibliography

Abbas, A.E. (2009). A Kullback-Leibler View of Linear and Log-Linear Pools. *Decision Analysis, 6*(1), 25–37.

Arló-Costa, H., Hendricks, V.F., Van Benthem, J., Boensvang, H., & Rendsvig, R.K. (2016). *Readings in Formal Epistemology: Sourcebook.* Springer.

Arntzenius, F., Elga, A., & Hawthorne, J. (2004). Bayesianism, Infinite Decisions, and Binding. *Mind, 113*(450), 251–283.

Atkinson, D. & Peijnenburg, J. (2013). A consistent set of infinite-order probabilities. *International Journal of Approximate Reasoning, 54*(9), 1351–1360.

Bana, G. (2016). On the formal consistency of the Principal Principle. *Philosophy of Science, 83*(5), 988–1001.

Basseville, M. (2013). Divergence measures for statistical dataprocessing–an annotated bibliography. *Signal Processing, 93*(4), 621–633.

Basu, A., Shioya, H., & Park, C. (2011). *Statistical inference: the minimum distance approach.* Taylor & Francis.

Belnap, N. (2003). No-common-cause EPR-like funny business in branching space-times. *Philosophical Studies, 114*(3), 199–221.

Bickel, J.E. (2007). Some Comparisons among Quadratic, Spherical, and Logarithmic Scoring Rules. *Decision Analysis, 4*(2), 49–65.

Bradley, D.J. & Leitgeb, H. (2006). When betting odds and credences come apart: more worries for Dutch book arguments. *Analysis, 66*(290), 119–127.

Bradley, R. (2005). Radical probabilism and Bayesian conditioning. *Philosophy of Science, 72*(2), 342–364.

Brier, G.W. (1950). Verification of forecasts expressed in terms of probability. *Monthly Weather Review, 78*(1), 25–27.

Briggs, R. (2009). Distorted Reflection. *Philosophical Review, 118*(1), 59–85.

Childers, T. (2013). *Philosophy and Probability*. Oxford University Press.

Cohen, J. & Callender, C. (2009). A better best system account of lawhood. *Philosophical Studies, 145*(April), 1–34.

Cover, T.M. & Thomas, J.A. (2006). *Elements of information theory*. Wiley-Interscience.

Csiszár, I. (1991). Why Least Squares and Maximum Entropy? An Axiomatic Approach to Inference for Linear Inverse Problems. *The Annals of Statistics, 19*(4), 2032–2066.

Dardashti, R., Glynn, L., Thébault, K., & Frisch, M. (2014). Unsharp Humean Chances in Statistical Physics: A Reply to Beisbart. In: M. Galavotti, D. Dieks, W. Gonzalez, S. Hartmann, T. Uebel, M. Weber (eds), *New Directions in the Philosophy of Science. The Philosophy of Science in a European Perspective*, vol. 5, Springer, Cham, 531–542.

Dawid, A.P., Lauritzen, S., & Parry, M. (2012). Proper local scoring rules on discrete sample spaces. *Annals of Statistics, 40*(1), 593–608.

De Bona, G., Cozman, F. G., & Finger, M. (2015). Generalized probabilistic satisfiability through integer programming. *Journal of the Brazilian Computer Society, 21*(11), 1–14.

Diaconis, P., Holmes, S., & Montgomery, R. (2007). Dynamical Bias in the Coin Toss. *SIAM Review, 49*(2), 211–235.

Diaconis, P. & Zabell, S.L. (1982). Updating Subjective Probability. *Journal of the American Statistical Association, 77*(380), 822–830.

Douven, I. & Romeijn, J.W. (2011). A new resolution of the Judy Benjamin problem. *Mind, 120*, 637–670.

Drouet, I. (2011). Propensities and conditional probabilities. *International Journal of Approximate Reasoning, 52*(2), 153–165.

Dziurosz-Serafinowicz, P. (2015). Maximum Relative Entropy Updating and the Value of Learning. *Entropy, 17*(3), 1146–1164.

Dziurosz-Serafinowicz, P. (2016). *The Double Life of Probability. A Philosophical Study of Chance and Credence*. University of Groningen.

Eells, E. & Fitelson, B. (2002). Symmetries and Asymmetries in Evidential Support. *Philosophical Studies, 107*, 129–142.

Elga, A. (2000). Self-locating belief and the Sleeping Beauty problem. *Analysis, 60*(2), 143–147.

Eriksson, L. & Hájek, A. (2007). What are degrees of belief? *Studia Logica, 86*, 183–213.

Fagin, R. & Halpern, J.Y. (1994). Reasoning about knowledge and probability. *Journal of the ACM, 41*(2), 340–367.

Fallis, D. & Lewis, P.J. (2016). The Brier Rule Is not a Good Measure of Epistemic Utility (and Other Useful Facts about Epistemic Betterness). *Australasian Journal of Philosophy, 94*(3), 576–590.

de Finetti, B. (1937/1964). Foresight: Its logical laws, its subjective sources. In: H. Kyburg, H. Smokler (eds), *Studies in Subjective Probability*. John Wiley & Sons.

van Fraassen, B.C. (1981). A Problem for Relative Information Minimizers in Probability Kinematics. *British Journal for the Philosophy of Science, 32*(4), 375–379.

van Fraassen, B.C. (1984). Belief and the Will. *The Journal of Philosophy, 81*(5), 235–256.

Freedman, D.A. (2003). Notes on the Dutch Book Argument. Available at https://www.stat.berkeley.edu/~census/dutchdef.pdf (accessed on 19 IV 2018).

Frisch, M. (2014). Why Physics Can't Explain Everything. In: A. Wilson (ed.), *Chance and Temporal Asymmetry*. Oxford University Press.

Gaifman, H. (1986). A theory of higher order probabilities. In: J.Y. Halpern (ed.), *Theoretical Aspects of Reasoning About Knowledge. Proceedings of the 1986 conference*, 275–292.

Gaifman, H. (1988). A theory of higher order probabilities. In: B. Skyrms & W. Harper (eds), *Causation, chance and credence*, 191–219. Springer.

Gaifman, H. & Vasudevan, A. (2012). Deceptive updating and minimal information methods. *Synthese, 187*(1), 147–178.

Gardenfors, P. (1982). Imaging and Conditionalization. *The Journal of Philosophy*, 79(12), 747–760.

Gillies, D. (1972). Review: The Subjective Theory of Probability. *British Journal for the Philosophy of Science*, 23(2), 138–157.

Gillies, D. (2000). *Philosophical theories of probability*. Routledge.

Goosens, W.K. (1979). Alternative Axiomatizations of Elementary Probability Theory. *Notre Dame Journal of Formal Logic, XX*(1), 227–239.

Greaves, H. & Wallace, D. (2006). Justifying conditionalization: Conditionalization maximizes expected epistemic utility. *Mind, 115*(July), 607–631.

Grove, A.J. & Halpern, J.Y. (1997). Probability Update: Conditioning vs. Cross-Entropy. In: *UAI'97 Proceedings of the Thirteenth conference on Uncertainty in artificial intelligence*. 208–214. Morgan Kaufmann Publishers Inc.

Gyenis, B. (2014). Bayes rules all: On the equivalence of various forms of learning in a probabilistic setting. Available at http://philsci-archive.pitt.edu/11230/ (accessed on 19 IV 2018).

Gyenis, B. & Wroński, L. (2017). Is it the Principal Principle that implies the Principle of Indifference? In: G. Hofer-Szabó, L. Wroński (eds), *Making it Formally Explicit. Probability, Causality and Indeterminism*, 35–41. Vol. 6 of the European Studies in Philosophy of Science series. Springer.

Gyenis, Z. & Rédei, M. (2016). Measure theoretic analysis of consistency of the Principal Principle. *Philosophy of Science*, *83*(5), 972–987.

Gyenis, Z. & Rédei, M. (2017). A Principled Analysis of Consistency of an Abstract Principal Principle. In: G. Hofer-Szabó, L. Wroński (eds), *Making it Formally Explicit. Probability, Causality and Indeterminism*, 3–33. Vol. 6 of the European Studies in Philosophy of Science series. Springer.

Hájek, A. (2008). Arguments for–or against–probabilism. *British Journal for the Philosophy of Science*, *59*, 793–819.

Hall, N. (1994). Correcting the Guide to Objective Chance. *Mind*, *103*(412), 505–517.

Hawthorne, J., Landes, J., Wallmann, C., & Williamson, J. (2017). The Principal Principle Implies the Principle of Indifference. *British Journal for the Philosophy of Science*, *68*(1), 123–131.

Hedden, B. (2013). Incoherence without exploitability. *Noûs*, *47*(3), 482–495.

Hughes, N. (2017). Uniqueness, Rationality, and the Norm of Belief. *Erkenntnis*. DOI 10.1007/s10670-017-9947-6

Huisman, L.M. (2014). On Indeterminate Updating of Credences. *Philosophy of Science*, *81*(4), 537–557.

Humphreys, P. (2004). Some considerations on conditional chances. *British Journal for the Philosophy of Science*, *55*(4), 667–680.

Imkeller, P. (2002). Random times at which insiders can have free lunches. *Stochastics. An International Journal of Probability and Stochastic Processes*, *74*(1), 465–487.

Ismael, J. (2008). Raid! Dissolving the big, bad bug. *Noûs, 42*, 292–307.

Ismael, J. (2015). In Defense of IP: A Response to Pettigrew. *Noûs, 49*(1), 197–200.

Jeffrey, R. (1970). Review. *The Journal of Symbolic Logic, 35*(1), 124–127.

Joyce, J.M. (1998). A Nonpragmatic Vindication of Probabilism. *Philosophy of Science, 65*(4), 575–603.

Joyce, J.M. (2009). Accuracy and Coherence: Prospects for an Alethic Epistemology of Partial Belief. In: F. Huber, C. Schmidt-Petri (eds), *Degrees of belief*, 263–297.

Kallestrup, J. (2012). *Semantic externalism*. Routledge.

Kemeny, J.G. (1955). Fair Bets and Inductive Probabilities. *The Journal of Symbolic Logic, 20*(3), 263–273.

Landes, J. (2014). Strictly Proper Scoring Rules. Available at http://philsci-archive.pitt.edu/10696/ (accessed on 19 IV 2018).

Lange, M. (2000). Is Jeffrey Conditionalization Defective by Virtue of Being Non-Commutative? Remarks on the Sameness of Sensory Experiences. *Synthese, 123*, 393–403.

Lehman, R.S. (1955). On Confirmation and Rational Betting. *The Journal of Symbolic Logic, 20*(3), 251–262.

Leitgeb, H. (2014). The Stability Theory of Belief. *Philosophical Review, 123*(2), 131–171.

Leitgeb, H. & Pettigrew, R. (2010a). An Objective Justification of Bayesianism I: Measuring Inaccuracy. *Philosophy of Science, 77*(2), 201–235.

Leitgeb, H. & Pettigrew, R. (2010b). An Objective Justification of Bayesianism II: The Consequences of Minimizing Inaccuracy. *Philosophy of Science, 77*(2), 236–272.

Levinstein, B.A. (2012). Leitgeb and Pettigrew on Accuracy and Updating. *Philosophy of Science, 79*(3), 413–424.

Lewis, D. (1976). Probabilities of Conditionals and Conditional Probabilities. *The Philosophical Review, 85*(3), 297–315.

Lewis, D. (1979). Attitudes De Dicto and De Se. *Philosophical Review, 88*(4), 513–543.

Lewis, D. (1983). New work for a theory of universals. *Australasian Journal of Philosophy, 61*(4), 343–377.

Lewis, D. (1986a). A Subjectivist's Guide to Objective Chance (with postscripts). In: D. Lewis, *Philosophical Papers, Volume II*, 83–132.

Lewis, D. (1986b). *On the plurality of worlds*. B. Blackwell.

Lewis, D. (1994). Humean Supervenience Debugged. *Mind, 103*(412), 473–90.

Lewis, D. (1999). Why Conditionalize. In: D. Lewis, *Papers in Metaphysics and Epistemology: Volume 2*, 403–407. Cambridge University Press.

Lewis, P.J. & Fallis, D. (2016). Accuracy, conditionalization, and probabilism. Available at http://philsci-archive.pitt.edu/12517/ (accessed on 19 IV 2018).

Loewer, B. (2004). David Lewis' Humean Theory of Objective Chance. *Philosophy of Science, 71*(5), 1115–1125.

Lukits, S. (2014). The principle of maximum entropy and a problem in probability kinematics. *Synthese, 191*(7), 1409–1431.

MacFarlane, J. (2014). *Assessment sensitivity: relative truth and its applications*. Oxford University Press.

Malpass, A. & Wawer, J. (2012). A future for the thin red line. *Synthese, 188*(1), 117–142.

Marczyk, M. & Wroński, L. (2014). Completion of the Causal Completability Problem. *British Journal for the Philosophy of Science, 66*(2), 307–326.

Masterton, G. (2010). *Objective Chance. A Study in the Lewisian Tradition.* Filosofiska institutionen Uppsala.

Meacham, C.J. (2010). Two Mistakes Regarding the Principal Principle. *British Journal for the Philosophy of Science, 61*, 407–431.

Meyer, C.D. (2000). *Matrix Analysis and Applied Linear Algebra.* SIAM.

Miller, D. (1966). A Paradox of Information. *British Journal for the Philosophy of Science, 17*(1), 59–61.

Nielsen, F. & Nock, R. (2014). On the chi square and higher-order chi distances for approximating f-Divergences. *IEEE Signal Processing Letters, 21*(1), 10–13.

Nikeghbali, A. (2006). An essay on the general theory of stochastic processes. *Probability Surveys, 3*, 345–412.

Øhrstrøm, P. & Hasle, P. (2015). Future Contingents. In: E.N. Zalta (ed.), *The Stanford Encyclopedia of Philosophy (Winter 2015 Edition)*, https://plato.stanford.edu/archives/win2015/entries/future-contingents/ (accessed on 19 IV 2018).

Osherson, D. (2002). Order dependence and Jeffrey conditionalization. Available at https://www.princeton.edu/~osherson/papers/jeff3.pdf (accessed on 19 IV 2018).

Park, I. (2013). Simultaneous belief updates via successive Jeffrey conditionalization. *Synthese, 190*(16), 3511–3533.

Perea, A. (2009). A model of minimal probabilistic belief revision. *Theory and Decision, 67*, 163–222.

Pettigrew, R. (2015). What Chance-Credence Norms Should Not Be. *Noûs, 49*(1), 177–196.

Pettigrew, R. (2016). *Accuracy and the Laws of Credence.* Oxford University Press.

Pettit, P. (2001). Deliberative Democracy and the Discursive Dilemma. *Philosophical Issues, 11*, 268–299.

Predd, J.B., Seiringer, R., Lieb, E.H., Osherson, D., Poor, H.V., & Kulkarni, S.R. (2009). Probabilistic coherence and proper scoring rules. *IEEE Transactions on Information Theory, 55*, 4786–4792.

Ramsey, F.P. (1931). Truth and Probability. In: F. Ramsey, *The Foundations of Mathematics and Other Logical Essays*, 156–198.

Rees, O. (2010). Why Betting Odds and Credences Come Apart. Talk given at the LOFT 2010 9th Conference on Logic and the Foundations of Game and Decision Theory University of Toulouse (France), 5–7 July, 2010. Available at http://loft2010.csc.liv.ac.uk/papers/20.pdf (accessed on 19 IV 2018).

Roberts, J.T. (2001). Undermining Undermined: Why Humean Supervenience Never Needed to Be Debugged (Even If It's a Necessary Truth). *Philosophy of Science, 68*(3, Supplement: Proceedings of the 2000 Biennial Meeting of the Philosophy of Science Association. Part I: Contributed Papers), S98–S108.

Rosenthal, J.S. (2006). *A first look at rigorous probability theory (2nd edition)*. World Scientific Publishing Co. Pte. Ltd.

Schervish, M.J. & DeGroot, M.H. (2014). *Probability and Statistics*. Pearson.

Seidenfeld, T. (1986). Entropy and Uncertainty. *Philosophy of Science, 53*(4), 467–491.

Shimony, A. (1955). Coherence and the Axioms of Confirmation. *The Journal of Symbolic Logic, 20*(1), 1–28.

Stern, R. (2004). Does 'Ought' Imply 'Can'? And Did Kant Think It Does? *Utilitas, 16*(1), 42–61.

Teller, P. (1973). Conditionalization and Observation. *Synthese, 26*(2), 218–258.

Vineberg, S. (2016). Dutch Book Arguments. In: E.N. Zalta (ed.), *The Stanford Encyclopedia of Philosophy (Spring 2016 Edition)*, https://plato.stanford.edu/entries/dutch-book/ (accessed on 19 IV 2018).

Vranas, P.B.M. (2007). I Ought, Therefore I Can. *Philosophical Studies, 136*(2), 167–216.

Weisberg, J. (2009). Commutativity or Holism? A Dilemma for Conditionalizers. *British Journal for the Philosophy of Science, 60*(4), 793–812.

Williams, P. (1980). Bayesian Conditionalisation and the Principle of Minimum Information. *British Journal for the Philosophy of Science, 31*(2), 131–144.

Wroński, L. (2014). *Reichenbach's Paradise–Constructing the Realm of Probabilistic Common "Causes"*. De Gruyter Open.

Wroński, L. & Godziszewski, M.T. (2017). The Stubborn Non-probabilist—'Negation Incoherence' and a New Way to Block the Dutch Book Argument. In: A Baltag, J. Seligman, T. Yamada (eds), *Logic, Rationality, and Interaction. Proceedings of the 6th International Workshop, LORI 2017, Sapporo, Japan, September 11–14, 2017*, 256–267. Volume 10455 of the Lecture Notes in Computer Science book series. Springer.

COPY EDITOR
Karolina Wąsowska

PROOFREADER
Agnieszka Toczko-Rak

TYPESETTER
Leszek Wroński

Jagiellonian University Press
Editorial Offices: Michałowskiego 9/2, 31-126 Krakow
Phone: +48 12 663 23 80, +48 12 663 23 82, Fax: +48 12 663 23 83